HEINEMANN STUDIES IN BIOLOGY Number 7

Editorial Adviser: S. A. Barnett

The Behaviour of the Domestic Fowl

The Behaviour of the Domestic Fowl

by D. G. M. Wood-Gush, B.Sc., Dip. Animal Genetics, Ph.D

*Head of Ethology Section, Agricultural Research Council,
Poultry Research Centre, Edinburgh*

HEINEMANN EDUCATIONAL
BOOKS LTD: LONDON

Heinemann Educational Books Ltd
London   Edinburgh   Melbourne   Toronto
Johannesburg   Singapore   Auckland
Ibadan   Hong Kong   Nairobi   New Delhi

ISBN 0 435 62920 4
© D. G. M. Wood-Gush 1971
First published 1971

Cover photograph
Burmese Red Junglefowl, cock and hen
(*Bruce Coleman*)

Published by Heinemann Educational Books Ltd
48 Charles Street, London WIX 8AH
Printed in Great Britain by Butler and Tanner Ltd
Frome and London

290872

Preface

In recent years animal behaviour has become an increasingly popular field of study and its relevance to many other subjects is widely appreciated. In the fields of agriculture and veterinary science this appreciation, alas, has been relatively slow. However, with the growth of intensive husbandry systems and the desire for higher standards of animal welfare the subject has much to contribute to these fields, as well as to biology in general.

Amongst agricultural animals the domestic fowl is of increasing importance. In the United Kingdom in 1968 it provided about twenty-three per cent of the total income from agriculture live-stock, outstripping beef cattle, pigs, and sheep. It is also a common laboratory animal. This book therefore has been written not only as an aid to agriculturalists and veterinary surgeons but also for research workers in other fields who use the fowl, and who want to learn more about their experimental material. It is hoped that it will stimulate some animal behaviourists, who at present study wild birds, to expand their interests to the fowl as well, for its biology is better understood than that of most birds, and its importance is increasing rapidly in a hungry world.

It has not been written as an introductory textbook on animal behaviour, and those not conversant with the terminology of ethology or psychology are advised to use the book together with an elementary textbook on animal behaviour.

The author wishes to thank Dr A. J. Richardson and Mr I. J. G. Duncan for helpful criticisms, Miss H. A. Scott for assistance in checking the texts, and Dr S. A. Barnett for his help as Editor.

D. G. M. W.-G.

Contents

Contents

Introduction

As an agricultural animal the domestic fowl is of increasing economic importance in the highly-developed countries, and many of its husbandry problems concern its behaviour. It is also of unique interest for the student of the science of behaviour. Although the domestic fowl is dull in comparison to many wild birds, it is an excellent subject for certain laboratory studies on avian behaviour. Since its physiology, embryology, genetics, and pathology are better known than those of other birds, it is sometimes ideal material for testing theories derived from field studies. In common with most birds, and ourselves to a large extent, it relies mainly on sight and hearing for communication, and is mainly diurnal in its activity, so that experimentation is often easier than with mammals that have a highly developed sense of smell.

It is in its domestication and its biological fitness under a variety of husbandry systems, that perhaps its greatest scientific fascination lies. It is generally considered to be descended from one or all four species of Junglefowl belonging to the Genus *Gallus*. Some authorities consider it to have been evolved from the Burmese Red Junglefowl G. *gallus spadiceus* Bonneterre, but Hutt [117] suggests that the heavy asiatic breeds may have evolved from more than one species.

The four species of Gallus, which belong to the family Phasianidae, inhabit south east Asia. The Burmese Red Junglefowl is found from north eastern and central India, the extreme southern part of China, and south eastern Asia generally to Sumatra, Java, and Bali [42]. Beebe [24] states that the maximum altitude at which it is found is 2100 metres but that generally its range does not extend beyond 1500 metres. The vegetation of its habitat varies a great deal. Beebe states that it is very commonly found in bamboo forests, whereas Collias states that it is found in India in association with the sal tree (*Shorea robusta*) although bamboo may be present. The density of the plant cover varies over its range and in India the birds may be found in forests of high humidity, as well as in areas that are dry and open. The Gray Junglefowl (*G. sonnerati*, Temminck) is generally distributed in southern India [24] where it lives in areas covered with trees, shrubs, and *Euphorbia*. The Ceylon Junglefowl (*G. Lafayettii*, Lesson) is distributed throughout the island where it has a wide range of habitats. Beebe [24] maintains that in the lower slopes of the mountainous regions the species shows a marked seasonal migration which is connected with the fruiting of certain plants. According to Collias and Collias [42] the habitat of the Ceylon Junglefowl resembles that of the Burmese Red Junglefowl more than the habitat of the Gray Junglefowl which is the closest neighbour. The Javanese or Green Junglefowl (*G. varias*, Griffith), which Beebe suggested belongs to a different sub-genus from the other three species, inhabits the drier coastal belt of Java but may be found in the mountains up to an altitude of 700 metres.

Beebe has described the diets of the four species. The Red Junglefowl, which may live near cultivated land, is omnivorous but generally graminivorous. The larvae and eggs of termites are eaten. The Gray Junglefowl studied by him generally ate grass seeds, grain (from fields), fruit, berries, and insects. Its main diet appeared to be the berries of the nilloo shrub (*Strobilanthes* sp.). The Ceylon Junglefowl studied by him also lived mainly off the nilloo shrub and had a similar diet. The Javanese Junglefowl lives on berries, termites, the leaves of *Lantana mixla*, insects, grass, and cactus fruit. Unlike the other three species it apparently does not eat much grain while on the fields but insects instead. Collias and Collias state that the main foods of the Red, Gray, and Ceylon Junglefowl are berries, and that the Red Junglefowl, which they studied more closely, ate the fruits of a small tree, *Ehretia laevis*.

The following details of the breeding of the four species have been given by Beebe [24]. The breeding season of the Red Junglefowl varies according to its position within the species range; in the sub-Himalayan part breeding occurs from February to the end of

May. While in the Malayan Peninsula breeding is from February until August. Five to eight eggs are usually laid but up to fourteen have been found in a single nest. Incubation lasts for twenty-one days. The Gray Junglefowl breeds from October to July while the Ceylon Junglefowl breeds all the year round but with a peak between February and May. The Javanese Junglefowl breeds during the dry season from June to November and lays six to twelve eggs. The colour of the eggs varies from white to café au lait in the Gray and Ceylon Junglefowl. The choice of nest site and type of nest appears to be fairly similar in all species. The nests are commonly found in thickets next to tree trunks, stumps, or clumps of grass. The nest is sometimes merely a hollow and in other cases is lined with grass and leaves. Occasionally nests are found off the ground on tree stumps and Beebe describes one Ceylon Junglefowl nest occupying a disused nest high in a tree – an act which scarcely seems adaptive. However, a number of nests of the Ceylon, Gray, and Javanese Junglefowl are reported as being several feet off the ground. The males of all the species may be either monogamous or polygamous, and in none does the male share in incubation.

Among Red Junglefowl unisexual groups or single cocks were common [42]; in fact single cocks were seen on thirty-five occasions and several cocks together another nine times. Flock size appeared to be about five birds where the birds roosted. Beebe's comments on the other species indicates that their social structure is similar to that of the Red Junglefowl. However, in the later chapters we discuss the behaviour of the Red Junglefowl under laboratory conditions more fully and in relation to the domesticated member of the genus.

Exactly how long the fowl has been domesticated is uncertain. The first records of it appear in Asia, and certainly by 2500 BC it was present in India in the Indus Valley cultures [44, 254]. From India it moved to Persia at a very early date and according to Coltherd [44] it was known in Mesopotamia about 2200 BC. Lowe [152] gives evidence of its presence in Crete in 1550 BC. It appears sporadically in inscriptions in Ancient Egypt but was probably common there only after the Persian invasion in about 525 BC.

Peters [176] thinks that the fowl entered Europe by two routes: through the Dardanelles to the Aegean and through Scythia to the Teutons and Celts.

Many of the ancient civilizations were deeply impressed by the crowing of the cock. In Zoroastrian literature of the Kianian period (2000 to 700 BC) the cock was known as the Herald of the Dawn. To what extent there was unconscious selection for frequency of crowing is an interesting question. Sauer [194] believes that interest in cock fighting was more responsible for the spread of the fowl than its use as food, and in all likelihood there was selection for good fighting cocks fairly soon after its domestication.

It is impossible to say when the emphasis turned to fowls as a source of food, but a record in Egypt, during the reign of Thutmose III (circa 1501 to 1447 BC), mentions a bird that 'brings forth [or bears] every day'. By Roman times the poultry industry in the Graeco-Roman area was fairly complex. Columella [72] writes about the development of specialized breeds. According to him the Greeks bred their birds for fighting qualities, while the Romans were primarily interested in birds that gave the farmer a good return, and he mentions several breeds. Pliny [181] writes that there were birds that laid daily; hence, even allowing for poetic licence, by that time there were evidently some birds comparable with modern birds.

During its domestication the fowl has then meant different things to different people, and different traits must have undergone unconscious selection. Even today, with the geneticist's attention rivetted to production records, many behavioural traits are probably under heavy unconscious selection pressure. Hence we have an animal with prodigious productivity but which still has some traits which seem to have lost all survival value under modern husbandry conditions. At the same time, there are many agriculturally valuable stocks which have lost traits that would be essential for survival in the wild. By comparing domestic fowls with other gallinaceous species, in particular the Burmese Red Junglefowl, we can assess some of the seemingly non-adaptive behaviour we find in domestic stocks and estimate how other traits have been altered by domestication. Furthermore we have the extremely good fortune of having a population of domestic fowl that have been feral on an island off the coast of Australia for about forty years, and which is under predation of wild cats and sometimes shot at by visitors. They are being studied by G. McBride and his colleagues [160], so that in due course we shall have full information about the reverse of domestication.

1: Senses and Perception

Vision

The colour vision of the fowl has been investigated by many workers [249]. Lashley [143] successfully examined the fowl's spectrum and trained his birds to react to red (650 mμ), green (520 mμ), blue-green (500 mμ). He suggested that the fowl's spectrum is divided into at least five areas of widely different reactive values and that they correspond approximately to the conspicuous division of the human spectrum. Walls [239] states that the hen's colour system is trichromatic and probably identical with that of man.

The existence of simultaneous colour contrast in the fowl is suggested by the results of Revesz [184]. His birds were trained to take grain from a green card on a grey background. When given a grain on a contrast green that was produced by a grey card on a red background it was accepted, but grain on grey cards with other backgrounds was refused. However, this finding needs to be further tested.

The presence of the Purkinje shift in fowls was suggested by Katz and Revesz (127). This has been confirmed by electro-physiological methods by Armington and Thiede [10] who compared the ERG spectral sensitivity curves for the light and the dark-adapted eye. The dark-adapted eye of the fowl responds to stimulation from wave lengths over the range 737 mμ to 420 mμ while the light-adapted eye gives strong responses over the range 636 mμ to 500 mμ.

The fowl has a range of accommodation of eight to twelve diopters [239]. Benner [25] stated and Dawkins [49] confirmed that shadow is a fowl's most important cue in depth discrimination. Hess [110] carried out an experiment to ascertain to what extent shading as a cue to depth depends on experience. He reared one group of chicks (controls) with light from above and another with light from below. The two groups were then given a critical test at the age of seven weeks by presenting each chick with a photograph of wheat grains, one half of which had been photographed with the light above them and the other half with the light from below. The first peck of each chick was noted. Ninety-five per cent of the control group pecked at the grains with the light from above while only 4·5 per cent of the experimental group (reared with the light from below) did so. Depth discrimination has also been

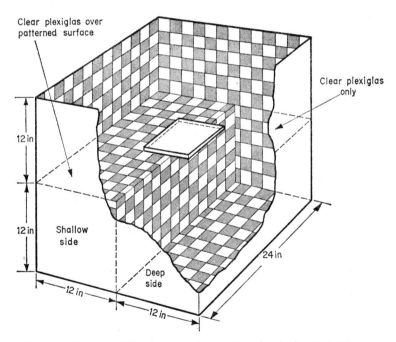

Figure 1. A visual cliff apparatus for testing depth discrimination in young chicks. The subject is placed in the box in the middle and if it can discriminate it will avoid getting out on to the glass.
(From Tallarico and Farrell [220].)

studied by means of the visual cliff [219, 197, 202, 220]. Typically a visual cliff consists of a glass sheet resting horizontally in a frame with a runway board extending across the middle of the glass (figure 1). On one side of this, a patterned surface is inserted directly under the glass and on the other side the clear glass rests well above a surface of the same pattern. The two halves of the apparatus are called the shallow and deep sides, respectively, because of their optical properties. In testing, animals are placed individually on the runway. If they consistently leave the runway and descend to the glass on the shallow side, it is concluded that they have responded to the depth cues from the patterned surfaces, for the two halves are identical in every other respect. Normally-reared chicks always prefer the shallow side of the cliff; in Tallarico's [219] study this preference was shown as early as three hours after hatching. Monocular chicks can make the discrimination, but the cues used for the discrimination are uncertain [197, 202].

The blinking response of the chick has also been used to study depth perception [69, 70]. At the age of three hours chicks responded to a feigned poke at the eye whether they had had experience of light before testing or not, and the rate of blinking to this stimulus was greater than to the stimulation from a fan extended so that the movement was across the line of vision rather than towards the eye. However, the blinking response of chicks that had been prematurely hatched by fifteen to forty-five minutes, and tested immediately after leaving the egg, was not directed towards a specific stimulus.

Hess [111], by testing chicks and adults in hoods fitted with prismatic lenses that would make objects appear closer to a binocular animal or laterally displaced to a monocular animal, concluded that the birds behaved as though they had binocular depth perception. The binocular field of thirteen birds was tested by Benner [25]. One eye of the fowl was covered with a shield, and over the head was placed a horizontal, graduated arc with its mid-point corresponding to the mid-point of the fowl's axes of sight. On the edge of the arc was placed a sliding clamp in such a manner that the angle between the mid-point and the clamp could be easily read on the arc. A piece of bait was tied to the clamp, and this was then

Figure 2. A chick fitted with hood having prisms that would make objects seem closer than they actually are to a binocular animal. (From Hess [111].)

gradually moved from the bird's blind side to its field of vision. As soon as the bird saw the bait it pecked at it and the angle between the position of the clamp at this time and the median plane was recorded. Benner estimated that the fowl has 26° of binocularity, but this needs to be tested with a more refined procedure.

In early work on form discrimination, Bingham [26] reported that, during attempts to establish the ability of the fowl to discriminate between forms, inversion of a triangle from the original position upset the fowl's discrimination. Munn [167] investigated some of the questions raised by Bingham's work. He was unable to train five four-month-old chickens to distinguish between a black cross and a black square, of equal area and brightness, both presented on white diamond-shaped backgrounds. When two different forms, each on a different background, were used he was able to train four animals to solve the problem. When the background of the positive stimulus – a white triangle – was inverted, three of the

chicks became inaccurate. Similarly when a white square was substituted for the white triangle they became confused. The changing of the negative stimulus did not affect their accuracy. The fourth bird, however, became confused when the negative stimulus was changed but was unaffected when the positive stimulus was changed. In a further experiment, two birds were trained to respond positively to a white diamond on a black circle and negatively to a white

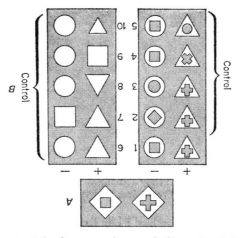

Figure 3. Some of the forms used to study form discrimination by the chick. Problem *A* was not mastered by the chicks. However, when the backgrounds were different as in *B* the Chicks learnt the discrimination. (From Munn [167].)

triangle on a similar black circle. Both circles were surrounded by white. In some control experiments one or both of the black circles were changed to triangles. In nine control experiments the entire background was white and one or both of the inner forms were changed. The chicks reacted positively to the point of light in the lower region of the figure (corresponding to the lowest angle of the diamond) and negatively to a straight or relatively straight line of light in the same relative position (corresponding to the base of the triangle). They were apparently uninfluenced by the backgrounds on which the white forms appeared. It seems therefore

that the chicks in Munn's (and Bingham's) experiments had responded not to the total configuration but to part of it.

Although the early work of Bingham, Munn, and others could
not demonstrate unequivocally that fowls could discriminate on
the grounds of form *per se*, preferences by fowls for certain forms
have been shown. Fantz [62] set out to determine the form of
preferences of newly-hatched chicks, tested in groups of twenty.
Each group was tested in four boxes each containing a set of four
objects; a sphere, an ellipsoid, a pyramid, and a star-shaped prism,
made of hardened tan plastic wood. The sizes of the objects varied
from box to box: the diameter of the sphere was 3 mm in the box
with the smallest objects and 7 mm in the box with the largest
objects.

Rounded forms were preferred to angular forms, but the possession of a rough texture decreased the popularity of the rounded
forms. In a control experiment in which the objects were covered
by transparent plastic the rounded forms still gained over ninety-
nine per cent of the pecks.

In a second experiment smaller groups of chicks were tested
during their first visual exposure with eight objects matched for
size and colour but tactile cues were excluded. The chicks were
one to three days of age at testing. Each box contained only two
of the forms but with a large and small specimen of each of the two.
(The small sphere was 3 mm diameter and the large one 4 mm.)
Smallness was slightly preferred to largeness but large round forms
received more pecks than small angular forms. These results were
corroborated in tests on single chicks.

In a further experiment the initial preference for rounded forms
was modified by experience of irregular food particles. The initial
preference of chicks for solid hemispheres as opposed to flat circles
was found by Dawkins [49] to increase owing, apparently, to a
waning in pecking at flat surfaces, for the waning could be arrested
by rearing chicks in boxes with corrugated sides.

Visual illusions have been reported by Revesz [185] and Winslow
[244]. In the former study the fowl's response was tested to the
illusion caused when segments of equal size are placed one above
the other. As shown in figure 4 the lower segment appears larger

to the human eye. Hens were first trained to take grain from the objectively smaller of two figures. When tested with two congruent segments the hen took grain from the subjectively smaller segment twelve out of thirteen times, and, on the next day, seven out of nine times. In control experiments the procedure was repeated with objectively different segments. Winslow [244] used a total of twenty-six birds and tested the responses to four types of visual illusions, but not all birds were tested with each illusion. Differences between birds were found, but generally the birds tested responded to the illusions as human subjects do. The birds were first trained to discriminate between a large and a small stimulus

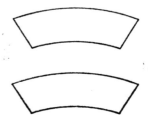

Figure 4. An optical illusion employed by Revesz [185] in testing the visual perception of the chicken.

that resembled the stimuli used in each visual illusion test with the objectively smaller stimulus as the positive one; most birds discriminated in favour of the subjectively smaller stimulus. The illusions (figure 5) were (i) a broken line which looks longer than a solid line of the same length, (ii) a horizontal line which appears shorter than a vertical line of equal length, (iii) a line with arrows at its end pointing outwards, which seems shorter than a line of equal length with arrows pointing inwards, and finally, (iv) the base of a rectangle of lesser height looks longer than the equally long base of a rectangle of greater height.

Although vision is the sensory modality that has received most attention we are still very ignorant about a number of factors: acuity, distance judgement, the threshold of response to light, the critical frequency, and the degree of individual variation are only some of the outstanding problems.

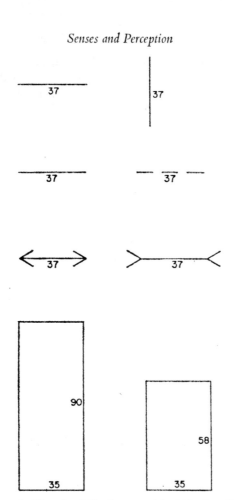

Figure 5. The visual illusions used by Winslow [244].

Taste

The demonstration of very fine preferences for certain solutions by chickens [124] suggested that the chicken has a much better developed sense of taste than had formerly been believed. Lindenmaier and Kare [148] reported the presence of taste buds in the fowl, but the appearance of these structures is very different from that of mammals, and indeed from the taste buds of pigeons [166].

Whether the structures described by Lindenmaier and Kare [148] are taste buds is yet to be verified. However, they describe the presence of a special type of epithelium together with many nerve endings beneath the mucosa. Acid and alkaline phosphatases and 3- and 5-nucleotidase are distributed throughout the basal layers of the mucosa of the fowl, whereas in the rabbit they are closely associated with the taste buds. Hence the receptors for taste in the fowl may be different even from those in the pigeon. Nevertheless, as we shall see, electrophysiological studies have shown that the fowl is able to taste.

The tongue of the fowl is innervated by two branches of the glossopharyngeal nerve [139]. The anterior one, called the lingual nerve, gives off branches not only to the pharynx but mainly to the anterior end of the tongue, the sides of the tongue, and to part of its posterior. The other branch, the laryngo-lingual nerve, goes to the larynx, and to the posterior and lateral parts of the tongue. No branches of the trigeminal nerve were found in the tongue. Recordings from both nerves responded to decline in tongue temperature, and so indicated the presence of cold receptors in the tongue. Furthermore the two nerves gave similar responses to chemical stimulation. The application of $0.5M$ NaCl solutions produced only moderate responses in some birds and gradually increasing sustained responses in others. Fifteen per cent sucrose and 0.06 per cent saccharin in Ringer's solution each produced a negligible effect compared with Ringer's solution alone. Fifteen per cent glycerine and fifteen per cent ethylene glycol both elicited prolonged effects. Responses were also obtained with $0.02M$ quinine hydrochloride and $0.2M$ acetic acid. NaCl, glycerine, quinine, and acetic acid solutions were all rejected by fowls given the opportunity to drink them. There was no marked behavioural response to sucrose at this concentration and no electrophysiological response (but, responses have been obtained to other concentrations [106]). Saccharin, however, is rejected behaviourally but evokes no electrophysiological response. Perhaps behavioural reactions depend on factors other than taste.

After this work Halpern [106] tested the response of the lingual nerve to twelve substances. Temperature was carefully controlled.

Responses were produced with $FeCl_3$ at or below 0·001M and with sucrose-octa-acetate at 0·0002M. Small but consistent responses were found with 1·0M sucrose and glucose and 0·5M xylose. Chickens were tested at several temperatures. Roughly, the birds could be divided into two groups: (i) those which gave as great a response to distilled water at 24°C as they did to the test substance at that temperature, and (ii) those which responded less to distilled water at 24°C than to the test substances. Responses to water, temperature, and various substances occurred only when the back

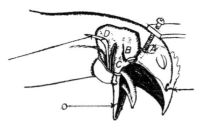

Figure 6. Schematic representation of the head and upper neck of the chicken preparation. A, ear bar; B, mandibular condyle (distal to excised section of mandible); C, portion of lingual branch of the glossopharangeal nerve from which recording is made; D, cornu (caudal segment) of the right hyoid bone, retracted ventro – and laterocaudally.
(From Halpern [106].)

of the tongue was stimulated, but mechanical stimulation was effective on both front and back. A calibrated hair which extended a force of 0·6 g over 0·08 mm² gave a clear response on the dorsum of the posterior tongue or the sides of the anterior tongue. Halpern, too, compared some of his findings with behavioural evidence: since $FeCl_3$ is rejected at a threshold close to 0·0001M, taste is probably the basis of its rejection. On the other hand sucrose-octa-acetate, which produces lingual nerve responses, is accepted as readily as water. As Halpern [107] points out, post-ingestial factors such as interoceptors, metabolic, and osmotic effects can overwhelm the simple relationship between gustatory neural responses and taste preferences determined by afferent information.

Hearing and other Perceptual senses

The range of hearing of the fowl is uncertain. Inferences might be drawn from experiments with other species but, since the pigeon and chicken differ so markedly in their taste mechanisms, such arguments must be accepted with reservation. Bremond [27] states that birds can hear sounds below 50 Hz and above 15 000 to 20 000 Hz. They may use the vibratory sense for perception of lower frequencies, and the bullfinch, *Pyrrhula pyrrhula*, is known to be superior to man in this sense. A bird can localize the direction of an auditory stimulus in three ways: comparison by the two ears of (i) the phase difference, (ii) the time difference, and (iii) intensity difference. The bird probably uses all three means at any one time but theoretically the first two methods are likely to be difficult in small species, and therefore also in the domestic chick.

The sense of smell of the chicken is believed to be very poor, but the olfactory nerve of the fowl shows electrophysiological responses to odorants, and under anaesthetic the fowl gives respiratory responses to amyl acetate at a concentration which is considered unlikely to involve the trigeminal nerve (Tucker [229]). However, the role of olfactory stimuli in the behaviour of the fowl has not been investigated. The pain and tactile senses have been neglected experimentally. Several workers have described tactile receptors in the fowl. Winkelmann and Myers [243] examined the roosters' skin taken from various regions. Both structural and histo-chemical examinations were made. Vater–Pacini end organs were found over the entire area of feathered and glabrous skin. A notable exception was the plantar surface of the foot where a type of Meissner corpuscle was found. The Vater–Pacini end organs were also found in the hard palate but not in the soft palate, tongue, the floor of the mouth or in the conjunctiva. In the perianal area the nerve endings found appeared to be encapsulated with a myelinated nerve passing through an arcuate structure. Winkelmann and Myers emphasize the general similarity between these cutaneous nerve endings and those found in mammals, and consider the differences to be relatively small in comparison. They also point out that the Meissner-form corpuscle found in the chicken foot is common among primates in which grasping is an important function.

Bantam hens in a cage which could be tilted so that gravitational and visual cues could be at variance were able to perceive horizontality by proprioceptive cues (Fantz [63]).

2: Communication

Fowls are highly social animals which, when allowed to do so, form a cohesive social structure. Communication is through visual displays and audible calls. In this chapter displays and calls are discussed in relation to the adult social unit. Their role in reproduction is discussed later. For the sake of convenience displays and calls are discussed separately, but this is an artificial division for displays are often accompanied by particular calls.

Displays

The waltz: the far wing is dropped and the cock advances sideways or circles round his opponent. The intensity varies: the wing may be fully or only slightly lowered and the distance moved varies. The neck may be extended or slightly retracted. When performed at its highest intensity the primaries touch the ground and the outer foot scratches against the primaries, producing a rasping sound.

Wing flapping is a very variable display. The cock may be at his full height or his head may be held flush with his back. The wings may be flapped noisily and fairly extended or when the second stance is adopted they are moved silently and are only slightly extended.

Tidbitting: the cock pecks at the ground perhaps giving food calls and scratching the ground.

Feather ruffling: the neck is stretched, the ruff is raised, the other feathers are ruffled and the whole body is shaken. In a breed like the Brown Leghorn, in which the ruff is golden in colour and the neck feathers long, it is to the human eye a most striking display.

Figure 7. A Brown Leghorn cock waltzing to a female.

Headshaking: in the most vigorous form the head is tilted to one side and vigorously shaken with circular movements.

Tailwagging: the tail is depressed until level with the back and moved rapidly from side to side in the horizontal plane.

Circling: the cock walks round the other bird with exaggeratedly high steps, watching it all the time. However, it is probably rare in the domestic cock for the outer wing is usually extended to some degree and in these instances it may be the start of waltzing.

Agonistic behaviour

Kruijt [136] has studied the development of fleeing and attack in the Burmese Red Junglefowl. Although the behaviour of the Junglefowl is not identical with that of the domestic fowl, his study provides a great deal of information which enables us to understand the development of display in the domestic fowl. The newly-hatched Red Jungle Fowl chick runs away from tactile and some auditory stimuli, but escape from visual stimuli develops only slowly. The auditory stimuli that elicit escape are sounds of long duration with little segmentation and lacking low frequencies. On day one the chick as part of its escape behaviour adopts the alert posture

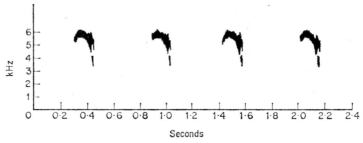

Figure 8. The shrill or distress call of the Burmese Red Junglefowl chick.
(From Kruijt [136].)

or squats or runs away. Freezing develops out of squatting but the chick freezes in the posture it happens to adopt at the moment of alarm. Two types of call are also given as part of escape behaviour: (i) distress calls (Figure 8) which are loud peeps repeated two and three times per second, and (ii) trill calls which are high trills lasting 0.5 second or less, consisting of five to twenty humps (figure 9).

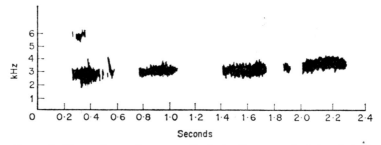

Figure 9. The trill or pleasure call of the Burmese Red Junglefowl chick.
(From Kruijt [136].)

Trills are seldom repeated like the distress call. From day one the chick gives the trill call when touched by another chick or human. At this age it squats to the clapping of hands, but on the second and third day the response to this sound becomes stronger, for the bird runs away and squats. The speed of running is greater on auditory stimulation than on tactile or visual stimulation. Escape as a response to escape responses of another chick does not develop until about the end of the second week.

Potential attack is less easy to follow than escape, but one of the earliest patterns is unorientated hopping which is seen on the second or third day. A chick becomes orientated towards other chicks at about seven days; frontal threatening may appear on the eighth and ninth days. Leaping at another chick is seen between days nine and twelve. However, in leaping, the legs are not thrown at the opponent as in adult fights, for kicking is seen only from day twenty-one. Aggressive pecking at the head appears on day ten but it is not seen in fights until three weeks of age. Hopping and fighting are highly correlated with general activity but not with the proximity of another bird. At this stage fighting leads to escape in the attacker and attacked alike, for the tactile or auditory stimulus from the other bird elicit escape.

From the age of three weeks in the Jungle Fowl, fighting becomes ambivalent and begins to contain escape patterns. Between days twenty-one and fifty irrelevant activities begin to appear during fighting. They are head-shaking, pecking, ground-pecking, incomplete ground-pecking, and head zig-zagging. Their prevalence is greater in fights without a decision than fights which end in a decision, although during this stage two-thirds of all fights end indecisively. At one to three months, ground-pecking, complete or incomplete, are mainly done by the winner, head-shaking and preening by the loser, and head zig-zagging by both. Between fifty and eighty days of age, when the peck-order becomes firmly established, attack and escape may both be sustained; most of the adult agonistic displays now appear.

The agonistic behaviour of the adult domestic cock has been described by Wood-Gush [250]. When one cock chases another he *struts*: the wings are trailed, the tail feathers and ruff may be raised and the legs are slightly bent. When a cock advances towards another he may *high-step*: the head is raised very high the chest is out, the tail is held high and the wings trail slightly. The cock's steps are very high but his path is usually obliquely towards the opponent and therefore may be indicative of a lower tendency to attack than strutting in which the cock faces his opponent. Very often, immediately before fighting, cocks assume the *fighting stance*: the head is held low, the legs are bent ready for springing, the neck is often stretched, the ruff raised, and the wings held out in a slightly trailing position away from the body ready for springing. When *retreating* the head is held up with the neck slightly retracted and the back slopes down to the tail which is depressed. The wings are held in position and do not trail, although sometimes one wing is raised and partly covers the back; the legs are bent. In *full retreat* the neck is stretched forward flush with the back, the tail may also be depressed, and the wings flapped as the animal runs away. None of these displays is absolute, for they often contain elements of other patterns. For example, the raised ruff, which accompanies attack, may be seen in full retreat.

Wood-Gush [250] studied the displays of fifty cocks in nearly three hundred encounters between males in neutral territory. Displays occurred in encounters between cocks (when the birds were apparently in conflict between attack and fleeing) but not in pure attacking or fleeing situations. In ninety-one waltzes analysed, the second cock fled from the waltzer or watched him from a fleeing stance in sixty-three cases, fought him or threatened him in sixteen, and took no action in twelve. In forty-six encounters the waltzer attacked or chased the other bird. It seems therefore to be indicative of a fairly high degree of aggression but the chance of no attack following it would indicate that the aggression is also checked. Kruijt is convinced that initially waltzing depends only on attack and escape, but that after the age of 120 days sex is also present as a third factor interacting with the other two.

Of ninety-seven wing flaps analysed by Wood-Gush twenty-eight were accompanied by steps towards the opponent and thirty by retreating. The remainder were performed standing. An analysis of the preceding and succeeding actions also revealed much ambivalence. Often it was followed by apparently irrelevant activities such as preening, tidbitting, head-shaking, bill-wiping or further wing flapping. It may occur when the cock is retreating but also happens when a cock springs at his opponent to attack him with his spurs. It probably serves to advertise the male's presence and in both the Red Jungle and domestic fowl males it is often followed by crowing. Tidbitting is also followed by a change from advancing or retreating. Often a cock changes his direction of movement in a single tidbitting action. Tidbitting may also be incomplete; this is very frequent when the cocks have assumed their fighting stance and their heads appear to be 'bobbing'. Sometimes elements from attacking or fleeing postures are superimposed. Kruijt states that tidbitting occurs in the Red Junglefowl from the age of 120 days.

Analysis of nineteen cases of feather-ruffling showed that the other bird took no obvious heed of it. In half, the performer changed from advancing to retreating. Often ruffling was followed by tidbitting and preening.

In these encounters also, head-shaking occurred in conjunction with tidbitting and preening whilst tail-wagging occurred in conjunction with wing-flapping and tidbitting. Preening occurred and was related to wing-flapping and tidbitting. Sometimes it was followed by a tendency for the cock to attack the other. Bill-wiping occurred in situations in which it was unlikely to have been mechanically functional and often was incomplete. In a third of the cases analysed there was an overt change from advancing or withdrawal but mostly it preceded or followed tidbitting.

McBride and his colleagues [160] have studied the position of the tail in relation to the activity of the bird in a group of feral domestic fowl inhabiting an island off the Coast of Queensland. When the birds are feeding in their own territory the tail is held in an intermediate position between the vertical and the horizontal or it is lowered from the horizontal. The dominant male with his group partly spreads his tail in the horizontal plane and his subordinates near him spread and raise theirs to an intermediate position. The position of the tail is perhaps used as a location system. However, with increasing caution the tail is raised and spreading occurs. Threat also involves spreading of the tail together with a lowering of the primary feathers. Minimum tail display, lowered and closed, indicates submission.

Waltzing, like tidbitting, occurs without calling. Foreman and Allee [71] have described the stances of hens in relation to the bird's

Figure 10. (a) The 'crouch' posture, (b) the 'semi-crouch' posture, (c) the 'deep crouch' posture, (d) the 'tall' posture, (e) the 'low' posture.
(From Foreman and Allee [71].)

success or failure in a paired combat. Three postures, crouch, semi-crouch, and deep crouch, are all connected with the winning of contests, and waltzes, when they occur follow semi-crouch and deep crouch postures. 'Tall' (posture), 'Low' (posture) and the sex crouch are associated with losing the contest. These stances are shown in figure 10. Kruijt describes another display which is common in both sexes of the Red Junglefowl: the *side display* may occur as a separate display or as a preliminary to waltzing. The bird moves in circles around the other bird, always keeping its side directed towards the

Figure 11. The side display by two adult Burmese Red Junglefowl. This display is similar to circling in the Brown Leghorn male, but generally very quickly breaks into waltzing.
(Kruijt [133].)

other. The oblique body posture and the position of tail and wings are the same as in waltzing except that the outer wing is kept folded. This is probably what is called *circling* in the domestic fowl [247]. However, it is probably rare in the domestic cock, for the outer wing is usually extended to some degree, and in a male is the start of waltzing. Conclusions on the side display in the domestic hen must await further study. The difference between the Red Jungle and domestic fowl males with regard to side display presents an interesting problem: has human selection led to change in the threshold of waltzing so that side display is now part of waltzing or, at the most, a rare display in the domestic cock?

The motivation of displays might be interpreted, in classical ethological terms, as an interplay of tendencies to flee, to attack and, frequently, to copulate. A detailed analysis of the elements of a display together with an examination of the immediately preceding and succeeding actions may suggest that these tendencies are simultaneously aroused. However, granted that displays may occur at changes of motivation, some displays may have become emancipated during evolution and come to possess their own motivational systems. Furthermore some displays may possess elements resembling attack and escape because of physiological factors common to themselves and one or both of these tendencies, without necessarily having evolved from an interplay between them. Electrical stimulation of the amygdala and of areas in the hypothalamus and mid-brain of the cat leads to threat displays which are a mixture of fleeing and attack elements [28]; it is not necessary to stimulate both attack-provoking areas and fleeing-provoking areas simultaneously to evoke threat behaviour. Von Holst and von Saint Paul [237] obtained threat behaviour from the domestic cock by stimulation of a single locus. It is not yet known to what extent a fowl's displays are emancipated from the neural processes which cause fleeing or attacking. Therefore it must be emphasised that in discussing the causation of a display in classical ethological terms, one is describing only one possible type of causation. Furthermore the motivation of a display may change during ontogeny.

Most of the male displays described here occur during courtship and some have value there as signals; some occur in parental situations and others during periods of stress (discussed in later sections). In the male to male encounters only waltzing appears to have a direct communicative function, as a warning signal, although wing-flapping in its most vigorous form may also serve as a warning. Other weak forms of wing-flapping may indicate submission. Nevertheless, most displays seem to have only small signal value in encounters between fowls of the same sex; the adaptive significance of much display may then lie in its effect on the performer. While some of the apparently irrelevant displays may be direct responses to the physiological or external stimuli normally connected with their occurrence [6], others may have 'cut-off' value as postulated by Chance [34]. To take two examples, in tidbitting the cock's notice will be removed temporarily from his opponent, likewise a preening cockerel has his eyes closed during preening. Such small shifts in attention could allow homeostatic physiological forces to act briefly, and herein may lie some of the adaptive value of many displays. It is to be hoped that studies in telemetry will shed some light on this possible type of function of the displays which was implied by Chance.

Calls of the fowl

The fowl's calls provide its main means of communication. As we have seen, most displays and calls probably function together. Few authors, however, have dealt with this relationship, and separate treatment is therefore necessary.

Many authors have tried to define calls of various types. Unfortunately only rarely, when audio-spectrographs are presented or when the stimulus situation is clear-cut, are the descriptions of much use. Even then there are short-comings. Sometimes no attention is paid to variation in calls between individuals from single breeds and between members of different breeds, nor have the stimulus situations been accurately described. For example, two authorities may describe as separate calls what may in fact be a single call which is released by two or more different stimuli.

The calls are listed below together with the name of the authority, and the stimulus situation which causes the fowl to elicit the call. Audio-spectrographs of some calls are shown in Figures 12–30. Inevitably different names have been given to the same call by different workers and these are all listed wherever cross reference has been possible. In those cases in which no audio-spectrograph has been presented a question mark is entered after the authorities' names. The breed of fowl used has not always been cited, and sometimes it has only been mentioned in relation to one aspect of an investigation, but the breeds used by some of the workers in this field are cited below:

Andrew [9]: unknown.
Baeumer [16]: possibly Orloff.

Collias and Joos [43]: New Hampshire chicks.
Adults–breed unknown.
Davis and Domm [48]: Breed unknown.
Guyomarc'h [103]: Rhode Island Red, White Wyandotte hybrid
chicks.
Wood-Gush [250]: Brown Leghorn Adults.

Chick calls

Calling begins whilst the chick is still in the shell but no comprehensive analysis of these pre-hatching calls has yet been made, and the list of chick calls refers exclusively to those of the hatched chick.

1. *Le cri d'appel du Poussin isolé, Guyomarc'h; distress call, Collias and Joos; peep, Andrew; Verlassenheits-weinen, Baeumer.* Guyomarc'h describes the call as being given in series at the rate of 2 to 2·5 calls per second when the chick is moved to a strange but light environment. If transferred to a dark or dim environment the rate increases to 3 to 4·5 calls per second and the series of calls is longer. The audio-spectrograph is shown in figure 12. Those shown by Guyomarc'h and Andrew for this call are all fairly similar but differ slightly in the range of frequencies covered which indicates differences in pitch.

2. *The short peep, Andrew.* The audio-spectrograph is shown in figure 13. The stimulus situation is not clearly defined but is probably the same as for the previous call. It resembles the distress call of Collias and Joos, and as Andrew states it is a transitional form.

3. *Le cri simple ou rythmique de plaisir, Guyomarc'h; twitter, Andrew; pleasure note, Collias and Joos; Stimmfühlungspiepen, Baeumer?* (figure 14). Guyomarc'h states that this is given when the chick rejoins its group after isolation and during transitory behaviour preceding the satisfaction of metabolic needs, for example feeding, drinking, and sleeping. It is given more intensely but more irregularly when the projected behaviour is in course. Collias and Joos state that it reflects a state of security.

4. *Circumflex call, Andrew;* The stimulus situation is not defined but is probably between those for calls nos. 12 and 14 (figure 15).

5. *Le cri complexe rythmique de sécurité, Guyomarc'h.* Given by an isolated chick when covered by the observer's hand or at the moment

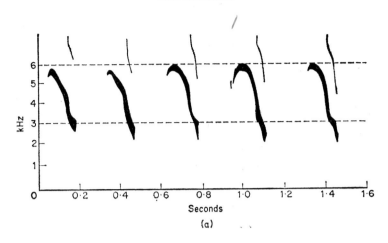

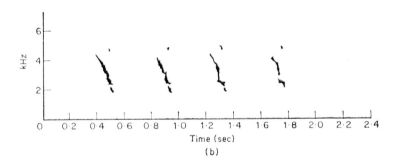

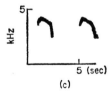

Figure 12. (a) Le cri d'appel du Poussin isolé; (b) Distress calls;
(c) Peeps.
((a) Guyomarc'h [103]; (b) Collias and Joos [43]; (c) Andrew [9].)

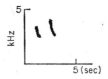

Figure 13. The short peep.
Andrew [9].)

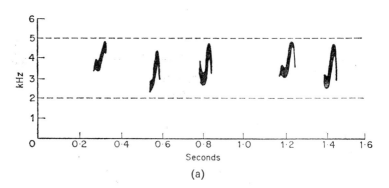

(a)

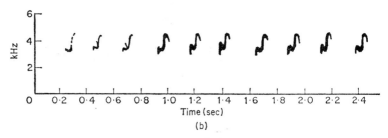

(b)

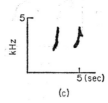

(c)

Figure 14. (a) Le cri simple ou rythmique de plaisir; (b) Pleasure notes;
(c) Twitter.
((a) Guyomarc'h [103]; (b) Collias and Joos [43]; (c) Andrew [9].)

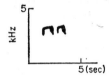

Figure 15. Circumflex call.
(Andrew [9].)

of covering up before sleep. The simple calls are repeated two to four times in a series, presented in two parts separated by a short silence. The first resembles a twitter and the second a peep. Guyomarc'h suggests that the call is indicative of short transitory phases of behaviour when the chick is establishing psychophysiological equilibrium (figure 16).

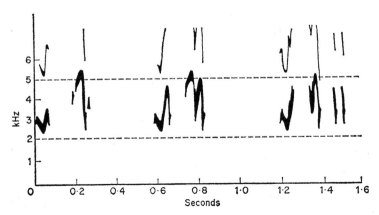

Figure 16. Le cri complex rythmique de sécurité.
(Guyomarc'h [103].)

6. *Le cri d'inconfort psychophysiologique modere (trille d'inconfort), Guyomarc'h; fear-trill, Collias and Joos; trill, Andrew; Schreck-triller der Kuken, Baeumer?* (figure 17). Released by contact with another bird or by the close approach of the hand of the observer or strange object. It is a trill of four to seven pulses of large amplitude emitted at the rate of thirty-two to thirty-seven pulses per second.

Guyomarc'h states that it occurs during transitory phases after a sudden change in the environment.

7. *Le cri d'effroi et le cri aigu de peur, Guyomarc'h; short squeal or fear note, Collias and Joos; shriek, Andrew; Angstgeschrei, Baeumer?*

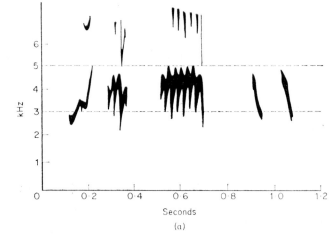

Seconds

(a)

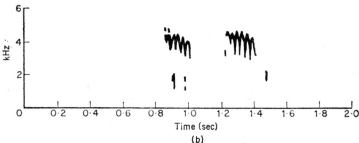

Time (sec)

(b)

Figure 17. (a) Le cri d'inconfort psycophysiologique modere; (b) Fear-trill.

((a) Guyomarc'h [103]; (b) Collias and Joos [43].)

Released by a direct act of aggression. Guyomarc'h states that it starts off as Le cri d'effroi and if the aggression persists, for example being grabbed, it evolves into le cri aigu de peur. It causes chicks of less than three weeks of age to regroup themselves by fleeing to a corner. He states that a week-old male chick treated with testosterone reacted aggressively to Le cri aigu de peur (figure 18).

8. *Le cri rythmique d'assoupissement, Guyomarc'h; Chevron call, Andrew; Kuken triller der schlafbereitschaft, Baeumer?* Guyomarc'h states that contact with other chicks is necessary for the elicitation of this call and darkness, if present, enhances its elicitation. It is

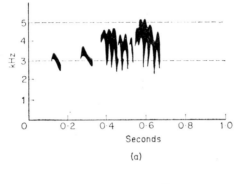

(a)

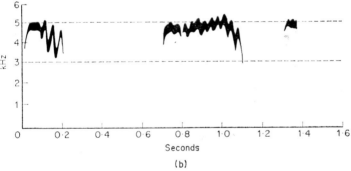

(b)

Figure 18. (a) Le cri d'effroi; (b) Le cri aigu de peur.
(Both from Guyomarc'h [103].)

characterized by two to three simple calls in a series of seven to ten calls per second. The modulation of frequency and amplitude of the calls is constant (figure 19). The audio-spectrographs of Guyomarc'h and Andrew for this call are very similar.

9. *Le cri rythmique de trouvaille, Guyomarc'h;* This is described as a rare call, but it was emitted by a forty-eight-hour-old chicken which

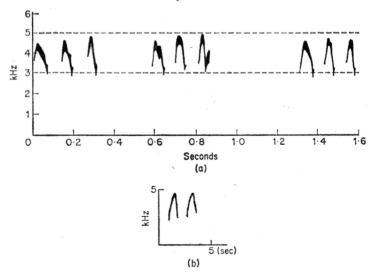

Figure 19. (a) Le cri rythmique d'assoupissement; (b) Chevron call.
((a) Guyomarc'h [103]; (b) Andrew [9].)

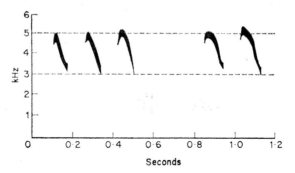

Figure 20. Le cri rythmique de deplacement rapide.
(Guyomarc'h [103].)

had found a food object. The calls may be punctuated by twitters.
10. *Le cri rythmique de deplacement rapide, Guyomarc'h* (figure 20).
This call is made by the chick when running and carrying a piece of
shaving. (Called 'food running' by Kruijt, see Chapter 5). The call
resembles Le cri d'appel du Poussin isolé in some respects.

11. *Le Trille d'alarme intense et autres trilles d'alarme, Guyomarc'h* (figure 21). Released by a loud sudden noise.

12. *Le cri rythmique d'expectative, Guyomarc'h; Gackeraufmerklaut*

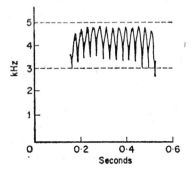

Figure 21. Le trille d'alarme intense.
(Guyomarc'h [103].)

Baeumer? (figure 22). Produced by a light noise during periods of inactivity, and is facilitated by darkness. During this call the animal stands in a particular way; the head is high and the chick scans the area. All other chicks on hearing the call adopt this posture.

The situations which give rise to peeping by chicks listed by Collias [41] are: isolated from companions, cold, hunger, thirst, pain, restraint, and approach of a large object. Andrew [9] has

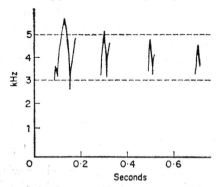

Figure 22. Le cri rythmique d'expectative.
(Guyomarc'h [103].)

suggested that twitters are elicited by moderate and relatively transitory stimulus contrast, and peeps by intense and very persistent stimulus contrast. He suggests three types of stimulus contrast. First, an intrinsic contrast with the background stimulation or the immediately preceding stimulation. Second, contrast due to persisting discrepancies between received stimulation and normal or preferred stimulation. Third, contrast acquired by stimuli when they become signals announcing some event such as feeding. He cites examples of chicks hatched in the dark giving twitters to mild electric shock.

Under this type of organization Andrew envisages that the calls of the chick form a single system of responses to specific classes of stimulus situations, although with learning many stimuli would come to be associated with certain types of primary reinforcement or punishment so that twitters and peeps would be emitted more specifically.

Calls of adult birds

1. *Crowing* (figure 23). Although this is the best known of all calls in the fowl its function is not very clear. It is mainly confined to adult males although occasionally laying hens may crow. Collias and Joos suggest that it advertises a territorial claim and may attract the female. Salomon, Lazorcheck, and Schein [192] found some evidence that perhaps supports the first of these suggestions: the crowing rate of a cockerel depends upon his position in the pecking order, and if placed in individual cages the crowing rates of socially inferior birds increases. Other explanations could account for this finding but it seems to support the idea of crowing being territorially motivated.

Konishi [133] states that in the audio-spectrogram it consists of four parts along the time axis. It contains many harmonic frequencies and there is a basic frequency-time pattern irrespective of individuals or of breeds. However, other differences can be found between individual birds.

Marler, Kreith, and Willis [155] studied crowing differences among young cockerels, from various stocks, that had been injected with testosterone propionate at the age of three days. They

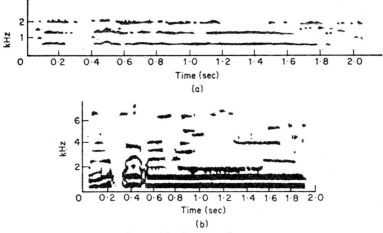

Figure 23: Crowing from,
((a) Collias and Joos [43]; (b) Konishi [133].)

compared the crows of closely related and non-related birds by means of seven criteria which included the presence or absence in the audio-spectrographs of overtones, distinct breaks in the temporal pattern, notes with a wide frequency spectrum, frequency oscillations of wide amplitude and even emphasis, frequency oscillations with emphasis on the peaks and/or troughs, and changes in the dominant frequency with time. Differences in the crows of chicks from the same strain were found and often individual differences were greater than the variability between strains. Only in highly inbred lines did the birds show less variability than the intra-strain differences. But in those lines even siblings differed from one another. Ontogenetic studies demonstrated that between the age of seven and twenty days the length of the crow increases, the maximum and minimum frequencies are reduced, and usually there is a reduction in the frequency spread. Overtones are rare in young birds' crows but are always found in older birds, and the terminal portion of the crow becomes drawn out and emphatic. Between forty-eight and eighty-two days of age there is a further increase in the duration of the call and the song is separated into three or so parts. The minimum frequency is reduced to below 500 cycles per second. The characteristic system of overtones becomes developed and the pattern of change of frequency with time becomes simplified. These characteristics were also found among untreated cockerels eighty-two-days-old.

2. *Warning calls.*

a. *Alarm note to hawks or aerial predator warning call, Collias and Joos; aerial alarm call, Konishi; Rähruf, Baeumer?* (figure 24). Described by Collias and Joos as a loud sustained and raucous scream that causes chicks to run and hide. It may be released by throwing a cloth

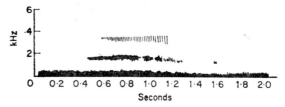

Figure 24. Aerial alarm call.
(Konishi [133].)

into the air, and has a wide spread of frequencies from 400 to 2700 cycles per second, combined with the presence of harmonic streaks. It is of long duration, and confined only to adult males. One of the audiospectrographs of Collias and Joos resembles that shown by Konishi.

b. *Fear squawks, Collias and Joos; distress calls and cries, Konishi; Angstgeschrei, Baeumer?* (figure 25). Given by a hen that is being held. According to Collias and Joos it resembles the alarm note to hawks in that it is sustained and has a mottled pattern. There are no frequencies below 400 cycles per second and the central part has an octave break. The drops in frequency resemble the distress calls of chicks. Konishi claims that these calls may assume many different frequency-time patterns and shows three different types of audiospectrographs, one of which resembles that of Collias and Joos. Baeumer lists another call, Wehlaute (pain call) which he says is similar to this call but not so shrill, and which occurs when the fowl has caught itself in something. In view of Konishi's observa-

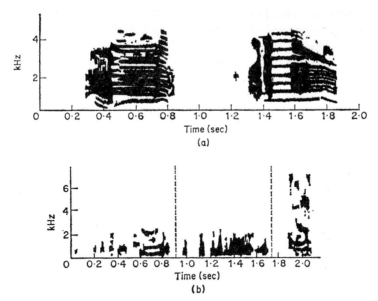

Figure 25. (a) Fear squawks; (b) Distress calls and cries.
((a) Collias and Joos [43]; (b) Konishi [133].)

tions it is possible that Baeumer's two calls might both be parts of the same continuum.

c. *Ground predator warning, Collias and Joos; warning calls – types 1 and 2, Konishi; Kleines Gackern, Baeumer?* (figure 26). According to Collias and Joos it is initially segmented and then ends on a sustained note of more than half a second long. It has no frequencies below 400 cycles per second; there is a good deal of harmonic structure with a pitch in the neighbourhood of the F that is half an octave above middle C. Konishi's audio-spectrograph for warning call Type 2 is similar to that for Collias and Joos' ground predator warning both being audio-spectrographs taken from males. Baeumer lists his Gackeraufmerklaut as a call given in response to a strange object on the ground. He describes it as go-go-*gook*-go, or *goo*-gogogock, from which it could be inferred that it is a combination of Konishi's Types 1 and 2 warning calls. Baeumer describes two other calls; one is the 'Grosses' Gackern in which the regular rhythmical calls of the 'Kleines' Gackern are followed by a longer accentuated sound which is given with the passing of danger. The second is the Legegackern which he considers to be the same as 'Grosses' Gackern but ritualized to a flock call and it may be given before egg-laying. These two may form part of Konishi's Type 1 and 2 warning calls, but until further research has been done the question must remain open. Konishi equates his Type 2 Warning

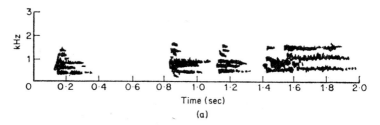

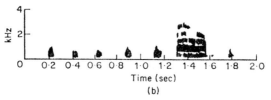

Figure 26. (a) Ground predator warning call; (b) Warning call Type 2 ((a) Collias and Joos [43]; (b) Konishi [133].)

Call and Baeumer's 'Grosses' Gackern with the hen's post-laying cackle.

d. *The threat sound of cocks, Collias and Joos; aggressive call, Konishi; 'Kollern' angeben, Baeumer?; whining call, sex call, Wood-Gush* (figure 27). Collias and Joos describe these calls as continuous calls with an emphasis on lower frequencies than in other warning signals. Konishi states that it sometimes also contains a series of pulse-like sounds which are more or less regularly spaced in time but they may be omitted completely. He also states that it is often associated with the waltz. The audio-spectrographs shown by Collias and Joos and Konishi for this call are dissimilar in some

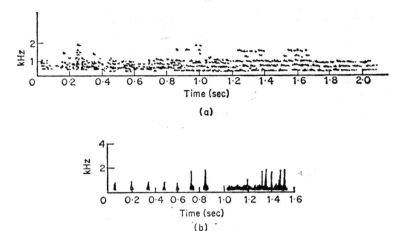

Figure 27. (a) Threat sound of cocks; (b) Aggressive call.
((a) Collias and Joos [43]; Konishi [133].)

respects although there is a sustained element of low frequencies in both audio-spectrographs.

3. *Tidbitting, Davis and Domm?; food call, Konishi; Futterlockrufe, Baeumer?* This call is usually produced by cocks and broody females but may sometimes be emitted by a normal hen. In courtship it is ritualized and in aggressive encounters it is emancipated from the feeding context. In both these situations food is not a necessary stimulus for its release. Konishi describes it as a series of pulse-like sounds with a wide frequency range delivered with a rather regular timing (figure 28).

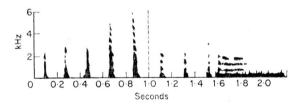

Figure 28. Food call.
(Konishi [133].)

4. *Alerting call, Konishi; Gackeraufmerklaut (Wachtlaut) Baeumer?* (figure 29). Konishi describes this call and equates it with Baeumer's Gackeraufmerklaut. He states that it resembles his warning call Type 2 but in contrast the pulsed elements usually follow a sound of longer duration. This sustained sound differs in the Type 2 warning call and the alert call for instead of the fundamental frequency remaining constant it rises and falls in the alert call. Like Type 1 warning call, the alert call is released in response to the presence of a passing animal or strange sound.

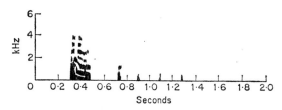

Figure 29. Alerting call.
(Konishi [133].)

5. *The laying call, Konishi; Gakeln, Baeumer?* This has been described by Konishi as consisting of sustained sounds with a low frequency range often including some harmonic frequencies. It may involve several component sounds with varying duration but with a total duration of more than four seconds. According to Baeumer it very occasionally is given by a defeated cock. Also given by the hen before egg laying and in other situations.

6. *The 'Ku' call of hens, Konishi; Rangordnungslaute freundlich, Baeumer?* According to Konishi it is a sustained low frequency sound usually without harmonic overtones. There is a tendency for the frequency to rise slightly towards the end of the sound. The duration is variable and there is a continuous gradation between this call and the laying call. Baeumer states that this call is connected with feeding or mating.

7. *Aggressive call of hens, Konishi; Rangordnungslaute Kampfandrohend, Baeumer?* The audio-spectrograph of this call is described by Konishi as being very similar to that of the sustained sound segment of the aggressive call of cocks. It has a very low frequency range

without harmonic frequencies. Its duration is also very variable. Hens produce this sound on meeting strange hens or cocks or even enemy animals according to Konishi.

A further ten calls apart from those given by the broody hen are described by Baeumer.

8. *Legegackern* (Cackling). Described by Baeumer as a ritualized flock call and given before laying.

9 and 10. *Nestabwehrlaute and Zischen im Nest*. These two are connected with defence of the nest. The first is described as prolonged, usually repeated, sounds given by laying or incubating hens. It varies from a rough growling to a shrill croaking and is therefore assumed to be a mixture of other calls. The Zischen im Nest is not described.

11. *Aufschrei nach dem gehackt werden*. Mono- or polysyllabic call which may either be loud or soft. It is given by a bird after being pecked.

12. *Beruhrungsabwehr*. A rough undulating sound, quiet or moderately strong. The individual sounds are half a second long. It is emitted when a bird gets pushed aside.

13. *Staubbad-abwehr*. Rarely emitted. The calls are repeated at intervals. Occurs when a bird is staking a claim to a dustbath.

14. *'Einsilbige' Aufmerkrufe*. Short sounds given in response to a flying bird or aeroplane. Two intensities are described.

15. *Girren*. Quiet elongated high pitched vibrating sounds which may last several seconds. Maybe simple or repeated. The bird emits them at rest and especially before going to sleep, or in response to disturbance by sounds or a visual stimulus in the distance.

16. *Rangordnungslaute leicht abneigend, beschwichtigend*. Short, quiet sounds of constant pitch given by females when they cannot get away. For example when sitting on the nest.

17. *Rangordungslaute herrisch*. Short varying sounds emitted by the subordinate bird in danger of being pecked.

18. *Nestgrundungslaute:* A long series of short quiet sounds as K–K–K–K with a slight additional vowel sound. Given by both male and female and occurs before the female has laid. This is possibly the tidbitting call associated with cornering.

19. *Pre-laying call, Wood-Gush*. This is given by the hen before

entering the nest and by pullets as they reach the age of lay (see figure 30). This may include Call 17.

The lengthy catalogue of calls shows the need of a full investigation for proper understanding: not only must audio-spectrographs be made for each supposed call but ranges of intensity should be cited, the stimulus situation very accurately described and details of the birds given. Only in this way will an objective list of calls be

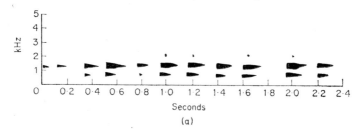

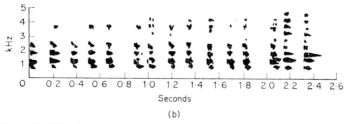

Figure **30.** The pre-laying call of the hen (a) when nesting behaviour is starting; (b) when nest-entry behaviour is further advanced. (From Wood-Gush.)

achieved. The problem of nomenclature too needs attention, not only to avoid proliferation but to reduce subjectivity in the description. For example, in Andrew's terminology 'twitter' and 'peep' are more objective than 'pleasure' and 'distress'.

Although the classification of calls needs much attention, certain principles regarding the coding of chicken calls are becoming apparent. Collias and Joos have shown that the sounds that attract young chicks have the following features: (i) repetitiveness, (ii) brief duration of the component notes, (iii) the presence of relatively

44

low frequencies of cycles per second. The sounds of a pencil tapping on wood or cardboard may copy all these qualities and in doing so, quieten the peeping of chicks and attract them. The common elements of the warning calls to chicks are: (i) relatively long duration, (ii) relatively little segmentation or repetition, (iii) absence or lack of emphasis on very low frequencies. Partial exceptions to these factors do exist, but the scraping of a wood chair gives an audio-spectrograph similar to the aerial predator call and has the same effect on the chicks' response.

In the feral population described by McBride and his co-workers, the adult birds are alerted by the chick-like sounds in calls not only against danger but when the cock is tidbitting or cornering (drawing the female's attention to a nest site). Calls may also have locating properties, as may the cackling of a laying hen (the Legegackern) which Baeumer has mentioned as being ritualized as a flock call. The calling of the Junglefowl hen after laying may also serve to draw predators away from the nest [242]. While moving through the vegetation these birds give calls (Legegackern?) and these become louder when two groups converge or when a hen is left behind; crowings too may serve as locating calls, for they are frequently given by the dominant male as he moves about his territory [160]. McBride further suggests that these crows cut down the chances of two groups meeting or serve to attract the members of one group.

Konishi points out that discrete calls are generally connected with specific external stimulus situations and that intra-call variation is correlated with the intensity of the stimulus situation. He suggests that calls which contain pulsed sounds or a multiple set of sounds can be varied in the rate at which the pulsed sounds or whole unit are given; the rate varies with the changes in the external stimulus. For pulse sounds a limit is reached at about four or five pulses per second. The bird may overcome this limitation by introducing a new element such as a sustained sound, as in Konishi's Type 1 warning call. On the other hand, calls that consist of a sustained sound are changed by increasing the duration of the call with increasing intensity of the stimulus situation. Further grading can be obtained by altering the interval between calls and their loudness.

3: Social Behaviour

Development of social behaviour

The social behaviour of the fowl may begin before hatching [138] but it is unlikely to be as striking as the social behaviour of the unhatched American bobwhite quail (*Colinus virginianus*) which leads to synchronous hatching. In this quail the appearance of a small hole in the shell occurs at variable ages in a group of eggs under artificial incubation but hatching is well synchronized [234] and this, it has been suggested, is connected with auditory interactions between embryos [234, 235]. However, Kuo [138] reported peeping in the unhatched chick as early as seventeen days, and responses to light and sound as occurring at seventeen and eighteen days respectively.

One of the first reactions of a newly-hatched chick is to press against something warm [214]. It may be asked whether the reinforcement so obtained is the foundation of all social behaviour in the fowl, and whether all social bonds are derived from other similar reinforcements obtained in the first few days after hatching.

Collias [41] states that the peeps (distress calls) of a newly-hatched chick away from the hen are caused by loss of warmth and loss of contact with the egg-shell. Ten chicks under a lamp, at a temperature of 37·8°C, averaged 100 peeps per chick for the first five minutes whilst another ten, hatched at a temperature between 25·6° and 27·8°C, averaged nearly 300 peeps per chick during the same period. If a hand was placed over the newly-hatched chick it gave very few peeps, whereas under the same conditions but without contact of the hand the chicks averaged 200–300 peeps per chick. His recently hatched chicks did not readily approach one another until after experiencing some minutes of bodily contact. Furthermore, the sight of a moving object did not decrease the number of peeps made by chicks until about one hour after hatching. Although a chick may approach a variety of stimuli without getting conventional reinforcement, such stimuli have not been considered outside the filial relationship; and the types of stimuli that are important in filial relationships may also be those that attract the chicks to one another in the usual type of artificial brooder. It may be asked whether the early responses of the chick to the parental model are any different from their responses to one another in the very early stages, when no natural or surrogate parent is present, or indeed whether the newly hatched chick is motivated by a general approach response which only later gives away to separate filial and social behaviour in which the chick has one set of responses to the dam and another to its siblings.

In considering the interplay of learning and heredity in the development of social behaviour, experiments in which chicks are reared under isolation have proved to have little value. The isolation is usually incomplete. If the growth of the chick is not to be impaired, the chick must be in light in which its own plumage will be visible to it and probably it will be able to see its own shadow; such experiences might direct it later to similarly shaped moving objects which possess plumage resembling its own.

Young chicks kept together soon interact. They are capable of co-ordinated walking within an hour of hatching and by three hours they show signs of having patterned vision [196]. Apart from Collias' observations [41] our knowledge of the chick's behaviour during its first few hours is very scanty, but Schaller and Emlen [196] tested several breeds for avoidance behaviour to a number of stimuli. White Leghorn-New Hampshire cross-bred chicks reared and tested in pairs responded more vigorously than singly-reared chicks tested alone from about twenty to 260 hours.

Smith [212] has given strong evidence of social facilitation among chicks during their second week. When they had to learn to run down a runway to get food, the presence of a trained chick enhanced the running speed of an untrained chick compared to that of an untrained chick running either alone or with another untrained chick. According to Collias [41] there is evidence of

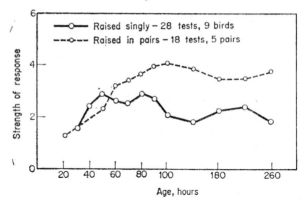

Figure 31. Comparison of the development of avoidance in birds raised and tested singly, with birds raised and tested in sibling pairs. (From Schaller and Emlen [196].)

leadership amongst chicks. Several groups each of ten to twelve chicks were placed in a narrow runway with heating lamps at each end, controlled by the experimenter. The chicks gathered under the lamp that was switched on. When it was switched off and the other switched on, some chicks responded sooner than others, and a few of these leaders sometimes repeatedly left the group under the warm lamp and went to a lagging chick in the cold, and this chick then followed the other back to warmth. Collias regards the 'leaders' as being less stimulus-bound than the others, but perhaps under these conditions many chicks do not relate warmth to the lamp but to 'keeping with the crowd' [249].

The peck-order

Chickens form what is known as the peck right type of status system, first described by Schjelderup-Ebbe [198]. In its simplest form in a group, say, of four birds one bird dominates the other three, the second dominates the other two, and the third is superior to the fourth bird. In many groups the order is not linear as in this example, and 'triangles' may occur in which bird A pecks birds B which pecks C which pecks A. There may also be 'square' relationships.

48

Schjelderup-Ebbe's observations that some flocks possess despots and 'Cinderellas' have been confirmed by many observers, and so has his observation that birds low in the scale generally attack their inferiors more than do birds higher up. In general, great contrasts of antipathy and toleration may be found between pairs of birds within a flock. In mixed flocks each sex forms a peck-order of its own. Guhl [89] reported evidence of a peck-order in a flock of ninety-six White Leghorn pullets although not all the pecking relationships were known. Under range conditions Fischel [65] found apparently loose groups, among birds away from the hen house, that were led by any hen, whereas among the feral fowls studied by McBride the dominant male led the flock.

The formation and duration of the peck-order

Guhl [91] has reported on the development of social organization among Plymouth Rock, White Leghorn and commercial hybrid chicks under quasi-agricultural conditions. Elements of social behaviour appeared in the following order: escape (avoidance), frolicking (an action in which individuals spontaneously run briefly with wings raised), sparring, aggressive pecking, avoidance of aggressive chicks, and fighting. Under these conditions escape was common from the third day after hatching and frolicking appeared during the first week, sometimes eliciting frolicking in other chicks. By the second week frolicking was seen to lead to sparring in which the chicks jumped up and down like adult birds but did not deliver any blows with their bills.

The earliest aggressive peck was noticed in the second week, but Guhl considered this to be exceptionally early. Avoidance of an aggressive chick by others was seen in the fifth week, but aggressive pecks were generally performed by the male chicks: between the second and sixth weeks they delivered 118 pecks to birds of both sexes, while the female chicks only gave thirteen pecks during this period. From seven to fifteen weeks ten male chicks delivered 3585 pecks during which time eight similarly aged female chicks gave 625 pecks. Two groups of young cockerels formed their peck-rights at seven and eight weeks while comparable female groups formed theirs between seven and nine weeks. In general, the chick's position in this type of ontogenetic peck-order depends upon its rate of development. Some groups may, however, form their peck-orders later, depending upon the size of the group. In mixed groups the peck-order divides into two uni-sexual groups at ten to fifteen weeks of age.

Guhl [89] assembled six male chicks that had been visually isolated until nine weeks old and these birds formed a peck-order as quickly as controls reared normally. Hence frolicking and bodily

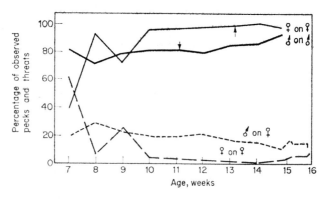

Figure 32. The development of unisexual pecking shown by the percentage of unisexual and heterosexual pecking occurring during the establishment and maturation of peck-orders. Arrows indicate the time at which the peck-orders were formed.
(From Guhl [91].)

contact with other chicks are evidently unnecessary for the formation of a peck-order. Guhl also compared the peck-order formed by eleven chicks reared together with one made by the same birds when reintroduced as strangers. These observations were compared with the findings from two sets of paired contests. For these the birds are at first isolated, and after a suitable time they are introduced into a neutral area for the contest. There was no evidence of a correlation between the original and final peck-orders. When the birds were introduced into a pen, correlations were found between (1) the number of birds pecked in the original peck-order and the number of contests won in the first series of paired encounters, (2) between the two sets of encounters, and (3) between the second set of paired encounters and the final peck-order. Guhl concluded that if the status of individual birds in a flock is to be assessed, more than one of the above-mentioned criteria should be used.

Factors determining status

The peck-order depends first upon factors that determine the birds' fighting skill and secondly upon determinants of the *status quo* such as memory and habit-strength.

Although some workers have used the two terms as synonyms, aggressiveness should not be equated with dominance: a dominant hen may be very tolerant while a hen of lower status is extremely aggressive. Defining aggression as the tendency to attack, Wood-Gush [251] recorded the aggressiveness of twenty-two adult cockerels and their positions in small peck-orders that were formed later. Initially the cockerels were kept in individual cages and each was pitted against every member of a panel of cockerels of fairly uniform agonistic behaviour. The birds were scored over a three-minute period for latency to start of fighting and whether they initiated the fight or not. No fights, once started, were allowed to continue. Each of the twenty two males had six such encounters at a rate of not more than one daily and the findings were then tested to ensure that the panel members had not changed their behaviour during the experiments. Later the twenty-two males were relegated randomly to one of three groups, each in a pen. After two weeks the status of the males was determined. Positive correlations were found between aggression scores but these were not correlated with success in the social hierarchies (figure 33).

Banks and Allee [17] studied the effect of flock size and composition on the frequency of pecking and threatening over a long period during which birds were studied in flocks of twenty-four, twelve, and six hens; each bird spent about three months in a flock of each size. Observations were made when the birds were feeding from a standard grain hopper which was introduced only for the observation period. Hens in the twelve-hen flocks pecked each other more than those in the other flocks. The six- and twelve-hen flocks also

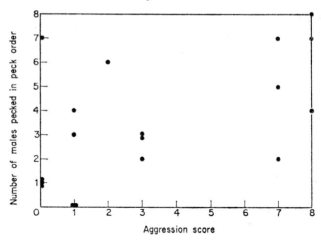

Figure 33. The relationship between a cockerel's score for aggressiveness (but not fighting-skill) and his position in a peck-order as determined from the records of 17 individuals.
(Adapted from Wood-Gush [251].)

showed more threat and avoidance behaviour than the twenty-four-hen group.

Unfortunately flock-size was compounded with the amount of feeding space per bird, and it is unjustifiable to attribute the amount of aggression found in this experiment to flock-size alone. For example, with a food hopper of constant size, many members of the twenty-four-hen flocks did not try to feed, and in the smallest flocks there was probably less competition for space. Therefore we must await further study of this problem.

King [129] recorded the pecks at one another of three groups of ten cockerels each. He used three conditions: (i) food scattered evenly over the floor, (ii) food in a dish of 45 cm diameter, (iii) food presented so that only one bird could feed at a time. The pecking frequency was negligible in the first condition and was highest in the third. Furthermore the peck-order broke down among some birds in the third condition, showing that attacks can be increased by competition. The implications of increased competition are discussed in Chapter 8. Competition, however, is not the only factor to increase attacks. Hale [104] investigated the effect of debeaking on the social hierarchy. Five debeaked hens formed a peck-order but had a considerably higher peck frequency than the control flock, owing to the lack of avoidance by the inferior birds.

Past experience influences aggressiveness and hence dominance. Potter [180] and Hale [105] studied the dominance relationships between different breeds. Some hens behaved in a characteristic manner to members of a particular breed rather than to their opponents as individuals. Evidently, once a social response has been developed towards a member of another breed, any member of that breed is likely later to evoke the same response. King [130] concluded from a study of thirty New Hampshire hens that once a dominance relationship is learnt it is not easily extinguished regardless of the initial fighting potential of the individual. Guhl [94] describes androgen-treated birds that failed to raise their status in their home pens but which were dominant and aggressive with strange birds. He suggests that the factors that lead to social inertia of this type also tend to promote stability in flocks [93].

Allee and his co-workers investigated the effect of various hormones on the aggressive behaviour of fowls. Allee, Collias, and Lutherman [1] examined the effect of testosterone propionate on the social position of birds. In a number of small flocks whose peck-orders were well known, several birds of low status received amounts varying from 0·25 mg to 1·25 mg daily for seven to eight weeks. Their behaviour was observed daily for three to five hours. Unfortunately a large number of birds died during these experiments and the numbers of completed observations were therefore small. Usually the larger the dose, the more rebellious the birds became. The revolts at first were against birds low in the peck-order, but later superior birds were overcome and high positions gained. Among controls receiving sesame oil injections daily, no social changes occurred. Long after treatment had ended and the overt effects of the hormone had vanished, the birds still retained their new positions. In a later paper [4] testosterone was tested on six breeds and strains. In a flock of ten Anconas the four lowest ranking birds were given daily injections of 1 mg in 0·5 ml of sesame oil daily for seventeen weeks. Some reversals of rank took place within seventeen days of the start of treatment and the treated birds ultimately attained the first three and fifth positions in the hierarchy. As had been found previously the birds retained their new ranks for the six months, long after their combs had regressed. In the other breeds fourteen birds improved their positions but two Brahmas, two New Hampshires, two Rhode Island Reds, and three White Leghorns from different strains failed to do so.

Nine birds from four flocks were injected with oestradiol, to test its influence on the peck-order [2]. Only two changes in social position were observed and both treated birds lost to social inferiors. A poulard receiving 1·0 mg daily for eighteen days crouched to the workers within three days of treatment while a control poulard struggled to escape. The next day, the treated poulard crouched to a cock while the control remained entirely unresponsive. Staged combats in neutral pens during the treatment indicated that the treated birds had lost aggressiveness. The authors tentatively conclude that treatment of over 0·3 mg daily for long periods decreases comb size and egg production, but they agree that these conclusions are based on too few cases.

The influence of thyroxin on status was investigated by Allee, Collias, and Beeman [3]. The birds in two flocks moulted slightly but the moulting did not affect the peck-order in these small flocks.

Guhl [87] set up two small experimental flocks of one cockerel, five capons, and six pullets, and one cockerel, four capons, and seven pullets, respectively. Five capons had very small combs, two had combs of intermediate size, and the remaining two had combs larger than those of the two normal males. On autopsy the four large combed capons were found to possess remnants of testes while the other five were without testicular material. The peck orders were determined early in life, for the birds had been reared together. With two exceptions the capons were dominant to the females. In one flock one female pecked four capons including one with an intermediate comb. In the other flock one small-combed capon was pecked by four females. These abnormalities may be due to the origin of the peck-order. It would be interesting to see the peck-order of a flock of capons and hens determined with adult birds, and the results of paired encounters before flock formation.

The position of the poulard in a peck-order is less definite. A normal hen has a left functional ovary and a rudimentary gonad on the other side. According to Domm [51, 52] many birds that had had their left ovary removed between the ages of six weeks and six months became indistinguishable from capons. Some attacked cocks introduced into their pens.

Figure 34. A picture of a unilaterally ovariectomized poulard showing its masculine appearance.
(From Domm [54].)

Autopsy revealed the presence of testicular tissue in the formerly rudimentary gonad and he reported spermatogenesis in some of these and other, similarly ovariectomized birds [52, 53]. Collias [39] states that when the ovaries were removed from six pullets selected from various positions in a flock of twelve birds, the treated birds fell below intact birds in status but retained the same ranks relative to one another. He does not state whether these birds had both ovaries removed or only one. If only one ovary was removed these results, in contrast to Domm's larger sample, could be due to incomplete ovariectomy or to ovarian regeneration or even to the length of observation, for the growth of testicular tissue is slow. Bilaterally ovariectomized pullets are unmasculine, inactive, and non-combative [52]. The presence of gonadal tissue in the hen appears to be necessary for aggressive pecking.

Domm and Davis [56] have studied the social status of intersexual birds. Oestrogen was injected into Brown Leghorn eggs on or

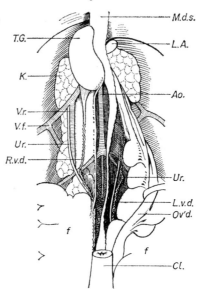

Figure 35. The right compensatory gonad, right and left convoluted vasa deferentia, and left oviduct. No trace of left gonad to be seen. L.A., left adrenal; Ao., aorta; Cl., cloaca; f., fat; T.G., testis-like right gonad; K., kidney; M.d.s., median dorsal mesentry; Ov'd., left oviduct (rudiment of right oviduct obscured); Ur., ureter; R.v.d., L.v.d., right and left vas deferens; V.f., femoral vein; V.r., renal vein.
(From Domm [54].)

before the fourth day of incubation. The resulting intersexual birds made a graded series in which the most feminine were nearly indistinguishable from normal females. Their gonads were ovotestes and commonly a left and often a right oviduct was present. The birds with the most masculine plumage were near the top of the peck-order and those with the most feminine were at the bottom. Birds that had received this treatment, regardless of their original sex, formed a single peck-order, by contrast with normal birds which form separate peck-orders, one of males, the other of females.

A number of studies have been carried out on the genetics of social dominance, although the authors have sometimes referred to this trait as aggressiveness. Komai, Craig, and Wearden [132] measured the social ranks of birds by counting all fights, pecks, threats, and avoidances. Percentile social ranks were calculated for each bird from the formula $(A + B)/2$, where $A =$ the percentage of the birds in the flock that the individual dominates, and $B = 100$ per cent minus the percentage of birds that dominate it. Three strains of White Leghorn and one each of Black Australorp, Rhode Island Red, and White Plymouth Rock were used. Social dominance was found to be partly heritable with mean intra-strain heritability of 0·30 and 0·34. Tindell and Craig [225] determined social dominance in the same way in different families in small flocks of White Leghorns and found negative correlations between social rank and sexual maturity. Guhl [92] selected two lines of White Leghorns that differed significantly in social dominance. Body weight and scores for social dominance were highly correlated except among the males of the low scoring line. Comb size showed no such correlation. In another study Craig, Ortman, and Guhl [45] carried out bi-directional selection of cockerels for five generations for social dominance and for scores in paired contests. Large strain differences for the two traits were produced in each of two breeds (White Leghorns and Rhode Island Reds). However, when the females from the selected strains were placed in flocks composed of females from the different strains their peck-order status appeared to be reduced compared to their ranks based on initial paired contests. For example, in one 'dominant' strain the females won eighty-seven per cent of the encounters with the females of a 'non-dominant' strain but when these females were housed together the 'dominant' strain dominated only sixty-five per cent of the 'non-dominant' strain. The authors point out dominance under the two types of conditions therefore, very likely, involve different variables.

Collias [38] analysed the factors correlated with fighting success in paired encounters, and some of his findings disagree with those of Craig, Ortman, and Guhl. To control for variables of sex, territory, and numbers present, he used only White Leghorn females, two at a time on neutral territory. He used the size of the comb as an index of the amount of male hormone present. The weight of the bird was deemed to give an indication of 'strength', 'impressiveness', and general health. Its rank in its home flock was used as a sign of the 'psychology of success'. The state of moult of the winning bird in each pair was also compared to the state of moult of the loser. Each of these factors was tested for correlation with success by means of the path coefficient method. Their order of importance was found to be: absence of moult, comb size, social rank, and weight. Collias, however, concluded that these factors as measured and used in the analysis accounted for only about fifty-six per cent of the factors involved, and suggested that more accurate indicators could possibly have given higher values for the paths, and thus decreased the importance of residual factors. Age appeared to be unimportant but experience in winning or losing was. He also suggested other factors such as differences in fighting skill, chance blows, differences in sensitivity to hormones, wildness, mild indisposition, past history of the individual, slight differences in handling, errors in judgement, and resemblance of opponent to a former despot.

Maintenance of the peck-order

The maintenance of a peck-order is generally assumed to depend upon the recognition of the individual bird by the others and upon their memory of their status relative to one another.

The means whereby birds recognize each other have proved difficult to discover. Schjelderup-Ebbe [199] disguised some birds and shortly afterwards returned them to their pens. The combs of some of the birds were turned to hang on the side opposite to the usual one, and these birds were attacked. The combs and wattles of other birds were stained violet, white, blue, black, yellow, red, and green. Those treated with blue and black were attacked, and so, to a smaller extent, were those dyed with green or violet. Yellow caused the hens to peck at it. Guhl [89] observed that when a bird, the comb of which had been removed, was returned to her pen after two days, together with a control pen-mate, she was attacked, but the control was accepted. This dubbed bird attacked her inferiors but submitted to her superiors. A flock of dubbed birds formed a peck-order, showing that factors other than the comb help in the recognition of the individual. Guhl and Ortman [90] extended Schjelderup-Ebbe's work on disguising birds from small, well-integrated flocks, and used as his criterion of recognition the acceptance of the bird without any challenges from members of the flock. Changes in the tail, wings, saddle back, and breast by the addition of ordinary or coloured feathers, or by denudation, evoked no challenges except where black or green had been used. Alterations to the head, neck, or comb were more effective in this respect, but duplicate disguises did not always give the same results. Deportment appears to be an important stimulus to attack, for a bird in a strange environment behaves in some manner detectable by the others and attracts them [90]. Similar observations were reported by Engelmann [61]. Many of these fifty-nine tests were inconclusive and the threshold of attack was very variable. A number of the attacks recorded were by inferior birds only, which raises an interesting point. Douglis [57] reported that when new birds were introduced into a flock, in fourteen per cent of the cases the alpha resident bird was the first to challenge the stranger, and in thirty-six per cent of the cases it was the second bird to do so. Therefore it is surprising that a considerable number of Guhl's disguised birds were challenged by inferior birds only. Guhl concludes that the recognition of the individual does not depend on a limited area of the body or on a single feature.

Hale [105] attempted to find out whether birds of one breed distinguished between individuals of another breed or whether they tended to treat all members of a breed alike. He investigated the problem in several ways: (i) by paired inter-breed contests between birds with previous experience of the other breed, (ii) by studying the peck-orders in small multi-breed flocks, (iii) by modifying the appearance of birds of one breed and studying its effect on the other breed that was penned with it. New Hampshires and Barred Plymouth Rock were used for experiments (i) and (iii) and again in the second experiment together with White Leghorns. Generally all the birds responded to all birds of the other breed in the same way and failed to discriminate between them individually. The few exceptions were individuals with highly variable experiences in paired contests, these tended subsequently to respond to members of the other breed on an individual basis. The discrimination may sometimes be very fine, for one strain of Black Australorps differentiated between two strains of White Leghorns [224]. These studies emphasize that the degree of variation in traits which fowls will tolerate in strange birds may have profound effects on breeding programmes, as discussed in the chapter dealing with sexual behaviour.

The duration of a bird's memory of its status after separation has been put at about two weeks [199]. However, this generalization needs testing, for memory of former flock mates may be influenced by the bird's age, the size of the flock and length of time they were flock-mates, and on the conditons in which the peck-order was formed [93]. Frequent reinforcement is necessary to ensure that a hen will remember her rank *vis à vis* her flock-mates [154]. Ten pairs of hens with established peck-orders were put into adjacent cages from which they could easily see and hear one another. Another ten pairs of hens were caged together for the same period. After three weeks the pairs of hens were released and no fighting was seen between members of a pair, but the experimental birds all immediately fought. Contact therefore appears to be necessary for the maintenance of recognition. Maier suggests that it is not necessarily pecking of the inferior hen by her superior that normally maintains the *status quo*, but a dominance-submission ritual which consists of head thrusts by the dominant hen and bowing movements by the submissive hen.

Smith and Hale [213] were able to change the dominance relationships in three groups of adult White Leghorn hens. There were four birds in each group, and the object was to change the peck-order from $A>B>C>D$ to $D>C>B>A$. Pairs of birds were introduced into a pen in which grain had been scattered over the floor, and the dominant bird was given an electric shock whenever it (i) threatened, pecked or attacked the other bird, (ii) tried to eat in the presence of its partner, or (iii) was attacked or pecked by the other bird. Each bird was allowed to eat in the pen in the absence of its partner so that no avoidance response was formed to food. In all groups the desired reversals were obtained. The reversal of the A birds required the least number of trials while the B and C birds needed more trials to learn their new ranks. The birds were also tested in another pen to ensure no effect due to the training pen. After the new rankings had been learnt, the birds were caged individually. Group 1 was then tested at two, four and six weeks after the end of training; Group 2 at three and five weeks, and Group 3 at five and nine weeks after training. The birds of all three groups retained the rank assumed through training. This is longer than the two to three weeks recall of peck-order status postulated by other authors, and the authors suggest that peck-orders developed through avoidance training may be more stable than those developed in a natural flock.

While, on the evidence of Smith and Hale, birds learn their status under some conditions, it may not be necessary to invoke learning to the extent previously believed in the formation and maintenance of a peck-order. Social dominance, we have seen, is partly genetically determined [132], and one may assume that birds differ in their aggressiveness. Casual observations on birds forming a peck-order show that many birds assume their relative positions without fighting, some birds always show flight, and others show flight to birds larger than themselves and to certain displays and plumage types. The subservience shown by such birds does suggest that only a trivial amount of learning is involved. Likewise other birds are seen to behave aggressively towards certain displays, plumage types, or to birds smaller than themselves, and again it does not seem that these dominance relationships need to involve the learning of differences between each individual. It is possible therefore that each bird may have a disposition to fight the birds showing some types of gait or display and/or structural features and to capitulate to others showing other types of carriage or display and/or structural features. Such differences in deportment and structure could form a scale to which the bird would respond according to its own condition. The discrimination need not necessarily be learnt for ethology is rich in examples of animals responding innately to releasers, and in the fowl Guhl and Ortman [90] showed that the presence of a large comb on an unaggressive bird caused others to flee from her. Under such a system fights would occur between birds whose positions on the scale were identical or nearly so. Variation in the threshold of birds to respond to others having features lower on the scale than themselves could explain the antipathy or tolerance sometimes found. Maier's results [154], it will be recalled, show the importance of physical contact and the possible necessity of frequent dominance-submissive rituals for the retention of naturally formed peck-order positions. This could be taken to indicate that the birds have to frequently see the dominance potential of the other bird relative to themselves before they respond, for the retention is not long. Many of the bird's relationships will be clear-cut for some flock mates will possess characteristics that may readily provoke flight in it for example, say, greater body size, and others will readily provoke threat from it. A peck-order based on a system like this would not involve a bird in learning its position relative to every other flock-mate, and retaining the information for long periods. The learning involved would be much simpler, and retention of individual relationships, as Maier's results have shown, would be much shorter than is presently believed. Learning would also be involved in cases such as those in which a bird loses to one or two birds in a strange area and then subsequently loses to all other birds in that area. Another possibility is that some relationships are learnt directly and that in small flocks learning may be important, but that in large flocks the other mechanism would be the more important.

McBride and his co-workers have shown that fowls in large populations are not distributed randomly in space [159] but that their spacing is based on visual stimulation of the other birds and is associated with the relative positions of the faces of other birds nearby. Evidence of territoriality has also been found [158], for when a wire partition separating two flocks was removed, birds crossing the old boundary were attacked and defeated so that two territories came into being. McBride's observations on the feral domestic fowl mentioned earlier also showed that the birds were organized into territories.

The effect of the peck-order

Since the peck-order forms the basis of all social behaviour in chickens its effect on the individual has been reported by several workers [39, 85, 89]. High ranking birds which deliver most

Figure 36. Alterations made to test for patterns of recognition.
Bottom right figure shows pen-mates avoiding a pullet after red
feathers had been added to her neck.
(After Guhl and Ortman [90].)

threats and win most fights also have priority for food, nests,
roosting places, and greater freedom of the pen. Guhl and Allee [84]
compared well-integrated flocks with an established peck-order to
flocks undergoing constant reorganization. They recorded the
amount of pecking, food consumption, body weight, egg production, size of comb, and social status. Generally the hens of organized

flocks pecked each other less than those of the experimental flocks. Food consumption was higher in the organized group than in either the experimental flocks or isolated birds. When feeding was restricted, the controls maintained body weight better and produced more eggs than birds in the experimental groups. Comb sizes, measured weekly, followed the same trends as egg production.

McBride [157] studied the production record and social status of Australorps kept in battery cages and in pens. All birds were scored for dominance by pitting each one singly against every member of a panel of six birds. No relationship was found between dominance and egg production in laying cages, but there was a non-linear relationship between dominance and egg-production among those housed on the floor. Birds kept in battery cages, however, are not always free from dominance relationships, particularly if they have to share feeding troughs. James and Foenander [120] studied 240 Australorp hens in cages in which two birds shared a trough. Each bird had three neighbours, one on each side and one at the rear, and therefore a possible peck-order position. The social rank of each bird in relation to her neighbours was determined by paired contests in a neutral pen. A peck-order, which was probably formed not long after the birds were put into the cages, was found. Birds low in the peck-order laid fewer eggs, varied more in egg number, and reached sexual maturity later than high-ranking birds. The authors suggest that the difference in sexual maturity was the main factor responsible for differences in production.

Tindell and Craig [224] also studied the effects of social competition on laying performance. The study covered two years. Three strains of White Leghorns with one strain each of Australorps, Rhode Island Reds, and White Plymouth Rocks were used. All birds were debeaked and housed in pens containing forty-eight or twenty-four birds with equal numbers from each strain. As in the study by James and Foenander [120] birds with higher social ranks matured earlier than low ranking birds. Social rank was positively and significantly correlated with body weight at five months of age, rate of feeding over 103 days and egg production over 119 days. From five to eight months gain in weight was negatively correlated with rank, but from the age of eight months there was no correlation between social rank and egg production. Some flocks were formed at five months of age and others at hatching, and as was mentioned earlier in the chapter, the age of assembly affected the relative status of a strain. The authors raise the question of social competition and genotypic potential, for if social dominance and production are genetically correlated it might be best to subject birds to an environment containing many chances of competition to allow more effective selection of both traits.

4: Reproductive Behaviour

Mating behaviour of males

Courtship

When placed with strange females or with familiar ones after a period of separation a normal cock usually begins his mating with courtship displays. Many of these displays were described in chapter 3. Waltzing, wing-flapping, tidbitting, feather-ruffling, head-shaking, tail-wagging, bill-wiping, preening, and strutting are observed during encounters both with other males and with females. Likewise the whining or sex call is common to both types of encounter. Waltzing, wing-flapping, and tidbitting are temporarily connected to approach, avoidance, and mating [250]. Waltzing and vigorous wing-flapping seem to intimidate the hen, for she mostly either moves away or crouches, rarely she responds aggressively. Tidbitting, if accompanied by tidbitting calls, on the other hand, tends to entice the hen towards the cock. The other displays mentioned above are connected with approach and avoidance but play no clear part in guiding the hen's behaviour.

Two displays, the Rear Approach and Cornering, are not found in aggressive situations. In the rear approach the cock approaches the hen, generally from behind, with his neck stretched and possibly his ruff raised. His deportment varies; he either holds himself high and moves with very high steps, or approaches more rapidly with less exaggeratedly high steps and the body kept lower. When the cocks are resident with hens this approach is common, but it is rarer among cocks put into pens for short periods. It is very often followed by mating.

Cornering by penned cocks has been described by Wood-Gush

[250]. The male runs to a corner, stamps his feet, lowers himself to the ground and usually gives the tidbitting call. Sometimes the sitting component is omitted and the male just stamps his feet in the corner. Although the male does not always go to a corner, the term cornering has been generally used. This action entices the females to the male. Its probable evolution makes it one of the most interesting displays in the domestic fowl. Among Junglefowl it is also a nest invitation display in which the cock guides the hen to a potential nesting site. McBride and his co-workers [160] describe a nest site display by the feral domestic cock which resembles cornering, although he does not give it any name. It may also occur during courtship of Junglefowl [147] but among domestic fowl living under semi-intensive husbandry conditions it no longer has any function in the nesting behaviour, at least in the breeds that have been closely studied.

In copulation the cock mounts the female from the rear, usually grasping her neck with his bill, and then makes treading movements with his legs for a few seconds and finally lowers his cloaca to contact that of the hen. Many matings are incomplete because the male stops after mounting and treading.

Development of male sexual behaviour

The development of male sexual behaviour has been studied by several workers. Andrew [7] reported juvenile copulation by two male chicks aged forty-eight hours, when the experimenter's hand, with fingers extended, was introduced into the chick's cage and thrust towards the bird. A number performed partial copulatory behaviour and at a slightly older age a small number of female chicks assumed the male role in the experimental situation. Testosterone injections were unnecessary to elicit this behaviour and even appeared to suppress it when injected into female chicks.

Juvenile tidbitting was also observed. Hence the neural mechanisms necessary for the performance of copulatory behaviour are present from an early age, but normally full copulation would not be seen because of lack of cooperation on the part of the other chicks. One performance of complete juvenile copulation delays further copulation for some time as with an adult, but the feed-back

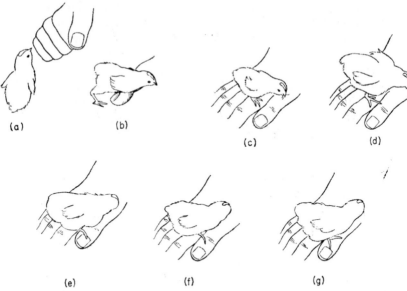

Figure 37. Juvenile copulation. (a) Erect, ready to leap up at hand, which is approaching. (b) Climbing on to hand immediately after it is lowered. Note chest is in contact with hand and bill is lowered to grasp at hand. (c) Bill grasps hand. (d) Wings are raised slightly, as pelvis lowering begins. (e) – (g) Treading, which caused the shift of foot position from (d) to (e), passes into quivering movements of the body.
(From Andrew [7].)

mechanisms are not necessarily the same, since the young birds cannot ejaculate, although constriction of the vas deferens is possible [7]. Nevertheless changes in the vas deferens would not account for females, adult or chicks, performing the male role. Andrew suggests that tactile stimulation from the ventral surface acts as a reinforcement for future mountings.

Andrew's chicks had been mainly kept in isolation from the time of hatching or from a very early age, and it is unlikely that juvenile copulation occurs in birds kept in groups, although attempted copulation may be common. The development of normal sexual behaviour in heterosexual or unisexual groups has not been studied as fully as it deserves. The best observations are those of Kruijt [136] on the Burmese Red Junglefowl and, although not identical to domestic fowl, their behaviour may act as a guide. Between the ages of thirty and eighty days copulatory behaviour may occur unmixed with aggression and escape, but no complete copulation was seen in this age group. Two behaviour patterns, treading (trampling) and copulatory sitting, which Kruijt considers to belong unambiguously to copulatory behaviour, occurred during this period. Copulatory sitting differs from sitting in other contexts in that the heels are turned out. Treading has already been described. During this age the rear approach is directed towards sitting chicks which do not allow themselves to be mounted. In tests with sitting models of chicks of equal age and of both sexes the experimental birds performed avoidance, aggressive, exploratory displays and the two incomplete copulatory actions mentioned above, in addition to the rear approach. Sometimes copulatory and aggressive behaviour were seen in the following combinations: rear approach with raised ruff and copulatory sitting with one or two of the following: raised ruff; leaping on to the model; wing flapping and vicious pecking at the model. At sixteen days of age manual massage of the abdominal muscles induced contraction of these muscles, bending of the legs, downward movement of the tail and opening of the cloaca during contraction, all of which are found in the adult male after the same treatment.

In each of three small groups composed of four cockerels and three to seven females the rear approach was released by the sitting of the lower-ranking birds of both sexes. Most of the rear approaches were by the lower-ranking males. In other groups containing five to eight males most copulatory responses were performed by the middle-ranking males. However, in the later groups the ratio of females to males was lower than in the others and the peck-orders were unstable. Furthermore, the small number of groups used makes the observations inconclusive. A very interesting finding was that low rank was connected with a high degree of response to manual massage, and Kruijt suggests that the amount of winning or losing of fights may lead to physiological changes which alter the readiness of the bird to perform copulatory contraction in response to tactile stimulation in the belly region. He suggests further that stimuli build up in the belly region and cause the male to mount and tread. However, the copulatory contractions found in females of this age group complicate this hypothesis. Kruijt suggests that the other copulatory components were mainly performed by the low-ranking males because in these males escape and aggression are more finely balanced than in high-ranking males and in fact much copulatory behaviour followed the comfort movements of head-shaking and preening. Nevertheless thirteen per cent of all copulatory responses at this age are unrelated in time to aggression or comfort movements. From 80 to 120 days the infantile calls change and adult calls are heard. Complete copulations do not occur, and copulatory behaviour does not change much in form, although more of it goes through to the treading stage. Mounting is now released by both standing as well as sitting birds of either sex and is no longer directed exclusively to birds of lower rank. Most of the mounting was found by Kruijt to be done by lower-ranking males.

From 120 days the males start to copulate successfully and this is often mixed with strong tendencies to attack. Kicking of the female during copulation was often witnessed by Kruijt but it is very rare among Brown Leghorns. Escape behaviour may also be present. Tidbitting and cornering appear in the cockerel's repertoire. From a detailed analysis of the temporal patterning of displays and examination of their components Kruijt concludes that the best working hypothesis to explain the causation of the main displays is that sexual, attack, and flight systems are continuously activated during encounters between a male and a female during this stage of the bird's life.

Several workers have studied the effect of early experience on the performance of mating and on the choice of a sexual partner. Wood-Gush [252] reared four Brown Leghorn cockerels each visually isolated from all other birds until they were six-and-a-half months old. Ten control cockerels were reared with other chicks in communal brooders to the age of eight weeks when they were placed in a pen together, until they were six-and-a-half months when they were placed in individual cages. In the pen some homo-

Figure **38**. Elements of copulatory and aggressive behaviour being performed by young Burmese Red Junglefowl males, aged between thirty and eighty days towards sitting chicks. (*a*), (*b*), and (*c*) copulatory; (*d*) copulatory attack; (*e*) and (*f*) attack; (*g*) and (*h*) exploratory. (From Kruijt [136].)

sexuality was observed. Their sole 'heterosexual' experience up to the test consisted of introducing a stuffed Brown Leghorn pullet in a crouching position into the pen when the males were fourteen weeks old. Several males attempted to tread it, but there was much jostling and none succeeded in completing the copulatory act. The mating trials were carried out in each of two pens of sexually experienced females. The males were tested singly for ten minutes in the late afternoon in each pen, and the order of release was randomised. Before the start of the mating tests, the 'isolated' males were given several hours each day in an empty pen where they could see no other birds, so that they could become acquainted with free movement and other conditions in the trial pens. No male was tested more than once a day. Three of the control males courted normally and copulated during their first test. The other seven attacked the hens for varying numbers of trials. Two remained aggressive for the eight trials, whilst one ceased to attack after the

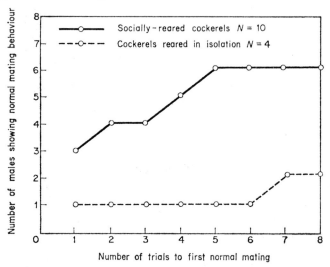

Figure 39. The effects of early social experience on the number of trials taken to first normal mating. Ten Brown Leghorn males reared in an all-male group until testing at six months of age and four Brown Leghorn males reared in visual isolation until six months of age. (From Wood-Gush [251].)

fifth trial but did not mate. Another copulated once but remained aggressive for all the eight trials. Kruijt [136] also reported aggressiveness by males raised in male groups until more than one year old. Three of the Brown Leghorn males that had tried to mount the dummy pullet when fourteen weeks old were amongst the most aggressive cockerels towards live females. One of the isolated males courted normally from his first trial and copulated on his third trial. Another copulated on his fifth trial after being aggressive. The other two behaved aggressively for all the eight trials. There appeared therefore to be no great difference between the controls and the males reared in isolation: many required a period of adjustment whilst a few mated normally from the start. Two of the males that had behaved homosexually, adopting the male role, scored the highest frequencies of copulation in the tests. Evidently males that perform normal coitus most readily also display copulatory behaviour in response to the widest variety of stimuli, for males which respond less readily a noticeable period of learning is necessary.

Guiton [99] examined the development of copulatory behaviour in two groups of males. Group *A* was kept communally from day one to day twenty and then in visual isolation from days twenty-one to forty-seven. Group *B* males were kept in visual isolation from day one to day forty-seven. From day four each male was given testosterone injections. On day forty-seven, after initial adjustment to the test arena over the previous few days, the males were tested with two objects, one a stuffed pullet in a crouching position and the other a yellow glove that had been worn by Guiton when attending and injecting the birds. While more Group *B* birds behaved sexually to the glove, a few 'which lacked any previous visual experience of a chicken did react sexually to the pullet on its first presentation'. Kruijt [136] made similar observations on the Junglefowl males that had been kept in isolation until six and nine months old; all copulated successfully within seven ten-minute encounters and matched the controls in this respect, but they did not display. Evidently normal copulatory behaviour is sometimes released by the normal sex object on the first trial.

Guiton's experiment emphasizes that for some males early experience is extremely important. For example more Group *B* males which had had no previous social behaviour behaved sexually towards the glove than did the socially experienced males of Group A. From day forty-seven to day fifty all males were kept in a large pen with pullets, and on day fifty-one they were tested again. The proportion of Group B males that reacted to the pullet now rose appreciably, but more still reacted to the glove than did Group A males. Fisher and Hale [68] report a similar attachment to the attendant by a cockerel that had been kept in isolation from hatching until at least eight months of age. Furthermore it attempted to mate with an overturned metal food trough. In an interesting experiment that needs repetition Vidal [233] found that six cockerels raised in pairs in visual isolation from other fowls from hatching until thirty to forty-five days of age, mated exclusively with a male model in a choice situation containing crouching models of both sexes matched for size, whereas six males reared together mated equally with both models.

Guiton kept his males in pens with females until six months of age, when they were caged individually and retested. At this stage they did not react sexually to the glove, and the group B males copulated as frequently with the hens as the Group A males in the test situation. The isolated males studied by Fisher and Hale [68] also copulated with the females after experience with them.

Guiton [97] studied two birds imprinted on to one of two cardboard boxes: at the age of fourteen months each responded sexually (tidbitting and waltzing) to the box on to which it had been imprinted, when given a choice between the two boxes. Between eight weeks and fourteen months they had been tested with the box only five or seven times. In a choice situation between a model hen and the imprinting box both birds vigorously attacked the model hen. Schoolland [201] describes cockerels mating with ducks in a pen containing eight cockerels, eight pullets, two ducks, and two drakes. He progressively decreased the number of cockerels and, as the number declined, so the frequency of cockerels treading also declined. From his description of this experiment it is unfortunately difficult to know whether the cockerels had been reared with ducklings from hatching. Evidently the early experience of the cockerel can guide its sexual behaviour, and early preferences can be changed.

Figure 40. A Brown Leghorn cock avoids the White Leghorn hen and waltzes to the Brown Leghorn hen in a choice situation.
(From Lill and Wood-Gush [146].)

Nevertheless, the form of the hen possesses high releasing value for sexual and agonistic behaviour regardless of the early experience of the male.

An attempt was made by Wood-Gush [258] to assess the sexual vigour of cockerels at an early age by their sexual responses under the influence of exogenous testosterone propionate, for adult males vary a great deal in their sexual vigour. Twenty-one male chicks were injected daily from the age of thirty-five days for ten days with 2 mg testosterone propionate. After the fifth injection the chicks were tested singly once daily for five days in an arena containing a freshly-killed chick in a crouching position. As adult males they were given ample sexual experience, and then tested under standard conditions. No correlation was found between the total number of matings in the adult tests and the dosage of testosterone propionate at which mounting and treading were seen in the juvenile tests, in fact two males that showed no sexual responses as juveniles were amongst the most sexually vigorous cockerels as adults. These results suggest that possibly the neural mechanisms involved in sexual behaviour mature at different rates in different animals and that for agricultural purposes, prepotent cockerels cannot be selected before they are mature.

Releasers for male mating behaviour

The valencies of the different parts of the hen as releasers for sexual behaviour by cockerels were investigated by Carbaugh, Schein and Hale [33]. Twenty sexually experienced White Leghorn males were tested singly with a complete model and at different times with a model lacking one or more parts of the body. The presence of the head is important for arousal of sexual behaviour, and it is possibly more effective when prone than when erect. The body alone is as effective as the head alone in arousing sexual activity but much more effective in releasing complete sexual behaviour. The body and tail presented together are also effective in arousing sexual behaviour but less effective in releasing full sexual behaviour than the body and head presented together. The head and tail together have a moderately high value for arousing sexual behaviour but did not release any complete sexual behaviour. The head is very important in ensuring correct orientation of copulatory behaviour. In another study with males that were reared in visual isolation rather unexpected findings were obtained with live hens,

Figure 41. Semen collection from a cockerel. Its legs are placed between the operator's knees. The right hand has finished the massage of the back, and the thumb and forefinger are in position to compress the erected copulatory organ.
(From Lake [140].)

for the males showed more mounting to standing or fleeing hens than to crouching ones which might have been expected to be more efficient releasers for sexual behaviour [68].

The frequency of mating in the fowl has a well defined diurnal rhythm [150, 173, 175, 208, 230]. Coitus is most frequent in the late afternoon. The greatest yields of semen and the greatest absolute numbers of spermatozoa occur at this time [139]. Control cockerels of the same age and strain as the semen donors were kept in pens with sexually receptive females; their greatest sexual activity coincided with the periods of the donors' maximum semen and sperm yield.

Variation in male sexual behaviour

Differences in the rate at which cocks copulate have been reported [85, 175 and 208]. Wood-Gush and Osborne [264] examined the mating frequencies of thirty males from six sire families. Differences were found not only in the rates at which the males mated, but also between the families. This indicates a genetical basis. However, the dams made no detectable contribution to the genetic variance. A negative correlation between comb size and mating frequency was found, suggesting that the differences in mating behaviour were not due to testosterone deficiencies in the low scoring families. Wood-Gush [256] selected two strains from two of these families for high and low rates of mating respectively and followed their behaviour over three generations. In the second generation the mean copulation rate and standard error per trial of the line with a high copulation rate was $11 \cdot 7 \pm 0 \cdot 79$, and that of the low line, $6 \cdot 25 \pm 0 \cdot 74$. When tested under standard conditions the two lines did not differ in the tendency to attack other males, crowing rate, acceptance of unusual food-stuffs or alarm calling. As the low-line males of the parental generation had had large combs, the males of both the first and second generation had been dubbed at the age of one day, in case the possession of large combs hindered development. In the second generation, however, although the two lines did not differ in body weight, the low line had larger wattles, hence the differences in secondary sexual characters found in the parental generation were still present. Several indirect measures of testosterone levels were used, but no differences were found and exogenous testosterone injections did not close the gap between the copulation rates of the two lines.

An attempt was made to find out whether adolescent experience was responsible for the differences between the two lines. In the first generation the low-line males had attacked the females more than the high-line males, and it was thought possible that this aggression was due to absence of sexual opportunities during adolescence. This hypothesis was tested by giving half the high-line and low-line males in the second generation some heterosexual experience from the age of thirteen weeks until the final tests at the age of six months. No differences in mating behaviour were found within lines that could be attributed to this early experience. All comparisons between the two lines suggest that the differences in copulation rate were due to central nervous factors rather than to endocrine or experiential differences.

The main difference, between lines apart from variation in mating frequency, was in semen production; in the second generation the low-line males produced 0.455 ± 0.043 ml and the high-line males 0.16 ± 0.46 ml from manual massage. However, when five males from each strain were massaged four times in sixty-five minutes only one low-strain male became exhausted and all the other males were yielding small amounts. An examination of the ejaculation pattern during copulation revealed another difference between the lines in the third generation. After each complete copulation the hen was caught and a vaginal smear was taken, and if no sperm were seen a small pipette was inserted into the oviduct to recover any semen that might be there. Seven out of thirteen first copulations in a test session on high-line males were without ejaculation whereas only one out of twelve of the low-line males first copulations were without ejaculation. This finding presents an interesting theoretical question as to which line was really the most highly sexually motivated line: the high line which copulated most or the low line which apparently required less stimulation to reach ejaculation? It also emphasizes the fact that sexual behaviour is not a unitary drive.

Siegal [206] carried out another selection experiment on the mating activity of cockerels. Using much larger numbers than

Wood-Gush he was able to show genetic rather than phenotypic correlations. Selection for high and low male mating activity was carried out for six generations. Mating was tested during eight ten-minute periods. A heritability of o·18 ± o·05 for increased mating was found and for decreased mating of o·31 ± o·11. Furthermore there was a negative genetic correlation between (i) mating rate and (ii) semen volume and sperm concentration.

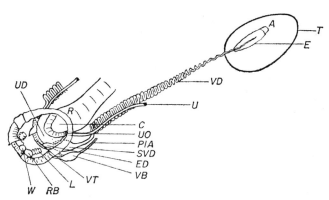

Figure 42. Diagram of the cloaca and reproductive tract of the male. The right side of the reproductive tract is represented *in situ* and the dorsal part of the cloaca is removed to expose the internal structure. Adrenal gland, A; Testis,T; Epididymal region, E; Distal portion of ureter, U; Opening of ureter into cloaca, UO; Rectum, R; Coprodaeum, C; Vas deferens, VD; Sac-like ending of vas deferens, SVD; Ejaculatory duct projecting into cloaca, ED; Fold of vascular tissue forming border of urodaeum and proctodaeum, VT; Urodaeum, UD; Internal vascular body (gefussreichen Korper), VB; Internal pudendal artery, PIA; Erectile structures in the proctodaeum, i.e. lymph fold, L; Round fold, RB; and White body, W.
(From Lake and El Jack [141].)

Copulation without ejaculation or with aspermic ejaculations is probably a common feature of the mating behaviour of the fowl for when copulations occur at frequencies as high as forty-one times a day [173] the male probably exhausts his supply of semen. When a device to collect semen was fitted over the cloaca of the male fourteen per cent of apparently complete matings were without apparent ejaculation, and such matings were commoner among males which mated frequently. The relationship between fertility and mating frequency is unlikely to be a simple one, and the highest fertility may be obtained from males which copulate at intermediate rates and in populations in which the ratio of females to males is not too high. Parker and Bernier [174] found an optimum ratio of six to seven males to 100 hens. The spatial distribution of the males is important, for if crowded together in small pens they form a peck-order which may affect the mating behaviour of all the males. In certain conditions the dominant cock does most of the mating and prevents the others from mating, although he may tolerate one inferior [85]. When the dominant cock is removed the other cocks do not tread often, even if they have previously been vigorous mating cocks. Guhl and his colleagues [85] conclude that the inferior cocks had been so trained that they were 'psychologically castrated'.

In a study of the mating frequency, aggressiveness and peck-order position of twenty-two cockerels, no correlation was found between mating frequency and number of males dominated in the peck-order nor between mating behaviour and several scores for aggressiveness [251]. Hence the dominant male is not necessarily the most active sexually and he might lower the fertility of the group by hindering other males from mating while he himself cannot. On the other hand, the presence of several males might enhance the potency of the dominant male, but this has not yet been investigated.

Mating behaviour of females

The generally passive role of the hen makes her behaviour more difficult to interpret than that of the male, but in well-integrated flocks mating usually occurs only when the female is ready. Rape is uncommon even if the mating is initiated by the cock. Often the female initiates coitus by crouching to the male. Occasionally a female mounts and treads with other hens whilst still in a reproductive state herself [86]. No detailed ontogenetic studies on the development of the sexual behaviour of the hen have been carried out. Fisher and Hale [68] reared some females under visual isolation

Figure 43. A Burmese Red Junglefowl hen crouching for a cock.
(From Kruijt [136].)

from each other, and as adult hens they crouched to human beings.
Their behaviour with sexually experienced males was not reported.
Junglefowl females reared under partial visual isolation crouched
to human beings and avoided the cocks [136].

Hens, like cocks, differ a great deal in the rates at which they
copulate [247, 253], but no comprehensive investigation has been
carried out to find out the causal mechanisms. The position of the
hen in the peck-order appears to be important; Guhl, Collias and
Allee [85] found a significant negative correlation between social
rank and the frequency of mating of White Leghorn hens in small
flocks. In other words, hens of higher rank mated less than their
inferiors. Similarly high-ranking hens were courted less than their
inferiors, and in some flocks the high-ranking hens crouched less
than the others. The rate at which a hen was courted was positively
correlated with the rate at which she was mated; sometimes there
was a positive correlation between the number of sex invitations
and number of matings.

Guhl [87] formed two heterosexual experimental flocks, each of
one cock and several capons and hens, to determine the influence
of the dominance of the males over the females upon the success of
mating. The normal males dominated the capons which, with a
few exceptions, dominated the hens. Most of the treading was done
by the normal males and the more masculine capons. Later, two of
the inferior capons were given oestrogen injections to increase their
sexual activity but not their aggressiveness. They then trod and
courted at high rates, but mostly with socially inferior females.
Superior hens often repulsed the small combed capons when they
tried to mate, and Guhl concluded that, although a male may mate
with superior females, social dominance of the male facilitates mating. He investigated further the influence of high social rank upon

the receptivity of the hen [88]. In his flocks the high-ranking hens crouched less than the lower- and middle-ranking hens, although hens of both high and low receptivity were found at all levels of the dominance orders. He then subdivided each flock into three levels of dominance and separated the three sub-flocks. The same males were used and rotated among the sub-flocks. The high-ranking hens now crouched more frequently than, or about as much as, the hens forming the lowest level. In two of the sub-flocks the middle and low-ranking hens crouched less after sub-flocking than before it.

Not only does mating vigour vary between hens but a single hen may vary from time to time. Wood-Gush [253] investigated the crouching rates of seventeen hens over periods of three to eleven months and compared these rates with the egg production of the hens. No correlation between rate of egg production and crouching rate emerged and, although the number of females investigated was not large, they represented a wide spectrum of reproductive activity, if a correlation between fecundity and fertility were general the sample would have revealed it.

Hormones and mating behaviour

The role of hormones in mating has been reviewed by Bastock [19], and by Wood-Gush [249] for the chicken. Although young chicks may spontaneously show male mating behaviour, testosterone certainly enhances it and can induce crowing [108]. Testosterone propionate is estimated to be thirteen times more effective

Figure 44. A fifteen-day-old male chick crowing after thirteen daily injections of 500 gamma testosterone propionate.
(From Hamilton [108].)

in inducing male mating behaviour by male chicks than alpha-oestradiol-benzoate [40]. Nevertheless oestrogen may be used to elicit male copulatory behaviour in capons [87]. Domm and Blivass [55] induced male copulatory behaviour in Brown Leghorn hens by subcutaneous implantations of testosterone. Several hens also waltzed and crowed. Injections of stilboestrol into two roosters caused them to become less aggressive and stop crowing, but they still copulated, though less vigorously than before [48].

Preferential mating

Upp [230] observed a flock of birds daily for two periods of three days each and later repeated another test on sixteen consecutive days. The cocks preferred certain hens but he could not see any behaviour by the preferred hens to account for this; hens that followed the cock around continuously were often neglected.

Preferential mating as measured by female solicitation of the male has been studied within an inbred line of Brown Leghorns [146, 247]. In both studies the females crouched at different rates to the individual males. In the earlier study the significant preference for one male over two others was probably based on his more vigorous courtship. In the second study, however, although there was some variation between the males in certain aspects of courtship (notably in cornering and tidbitting) there was no correlation between courtship behaviour and success in evoking female solicitation.

Lill and Wood-Gush [146] extended their study to compare interbreed preferences. Two true breeds, White Leghorn and Brown Leghorn, were used and two broiler strains originating from a four-way cross, White Leghorn × Cornish Game × Light Sussex × Rhode Island Red. These birds from the broiler stock therefore varied greatly in plumage colour. Both Brown and White Leghorn females were homogamous but the broiler females preferred the Brown Leghorn males (which closely resemble Red Junglefowl males). These female preferences did not appear to rest on courtship differences, for often the females began to discriminate in favour of, or against, a male immediately on his entry into the pen.

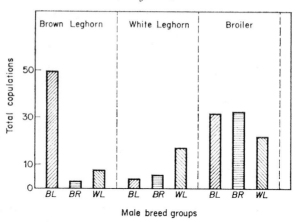

Figure 45. The solicitation rates of Brown Leghorn, White Leghorn, and Broiler males by females of the same stocks. *BL* = Brown Leghorn, *WL* = White Leghorn, and *BR* = Broiler. (From Lill and Wood-Gush [146].)

The females of three inbred Brown Leghorn lines solicited males of their own line more than males of the other two. The males of the three lines differed in appearance, behaviour, and pitch of vocalizations.

Male preferences were tested by releasing males into a pen containing two caged females equidistant from the point of entry. The number of displays to, and time spent near, a particular female were counted as being indicative of a male's preference towards her. Males that had been reared with females of their own breed were homogamous, but males reared in mixed breeds displayed only weak homogamy. This suggests that imprinting may help to determine adult preferences.

The existence of pronounced preferential or assortative mating among fowls suggests that behavioural isolating mechanisms can arise quickly at least under domestication. This is of practical importance, for assortative mating may lead to loss of fertility when cross breeding is attempted. Further difficulties with regard to gene flow could also arise from the social structure of fowls, since competition between males, by preventing certain males from mating, could exaggerate the effect of assortative mating. Furthermore, a cohesive status system could handicap newcomers until they became fully assimilated. Further detailed information is therefore needed on assimilation of immigrants into a population and on the type of social structure formed by fowls in the high population densities found in many deep litter houses.

Pre-laying and nesting behaviour

Before laying an egg the hen may perform a fairly involved behaviour pattern, the exact form of which depends largely on the environment. McBride, Parer and Foenander [160] give a description of the behaviour of feral fowls in the wild. When about to lay the hen utters the egg-call or pre-laying call, which is given with the bill open. The hen is usually joined by a male which accompanies her to a potential nesting site. The male may draw her attention to the site by giving a clucking-like call associated with tidbitting, followed by a dust bath sequence (cornering?) in or beside the potential nest site. The hen then enters and examines the site.

The domestic hen in the semi-intensive conditions of a pen gives the pre-laying call (figure 19) and may walk restlessly around the pen performing stereotyped movements, she then examines a number of nests before she finally enters one. The same nest or one of a group of adjacent nests is repeatedly used [248]; sometimes the same nest is used for more than one laying season [248]. In a battery cage pre-laying behaviour may be reduced to a number of stereotyped movements or to comfort movements [263]. Nevertheless in any one environment pre-laying behaviour is fairly constant. Some birds show persistent idiosyncrasies such as leaping against the wall of the pen or persistent calling.

Both the time spent 'examining' nests as measured from when the hen first puts her head into a nest until entry, and the time on the nest (the period from entry to oviposition), are related to the lag of the egg within the clutch [257]. The presence of the oviduct is not necessary for normal nesting, provided ovulation has occurred. The relationship of the two components of nesting behaviour to lag suggested that the post-ovulatory follicle might be implicated, and in fact, showed if this organ is disturbed by excision, ligation or the injection of cocaine, nesting is abolished or highly abnormal [75, 259]. Whether the post-ovulatory follicle acts by neural or by endocrine means has not yet been determined. The temporal relationships of this behaviour pattern suggest that a hormone is involved. Ovulation occurs some twenty-two hours before oviposition and, although the time of action of the post-ovulatory follicle is not known, the processes which lead to nesting at a particular time on one day are evidently triggered off on the previous day. It still has to be discovered whether ovulation is necessary, or whether processes in the CNS are all that is needed.

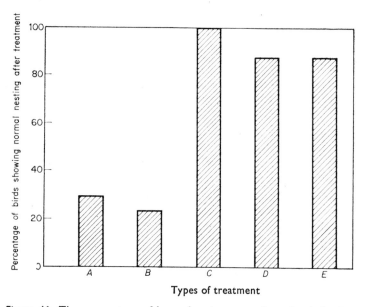

Figure 46. The percentage of hens showing normal nesting behaviour after various treatments either, to the post-ovulatory follicle shortly after ovulation or, to other parts of the ovary. A = Excision of the post-ovulatory follicle. B = Ligation of the post-ovulatory follicle. C = Manipulation only of the post-ovulatory follicle. D = The follicle, apparently next due to rupture, removed. E = Two immature follicles plus a small part of the ovarian wall removed.
(Adapted from Wood-Gush and Gilbert, [259].)

The exact stimuli from the nest that release entry and sitting have not yet been investigated, although they are of practical interest to the poultry industry. Detailed observations on hens in the nest have not been done, but one feature of interest is the occasional throwing of nesting material on to her back by the hen. This behaviour occurs among feral fowls when the hen is disturbed [160].

In the battery cage the hen may perform vacuum nest-building, making scooping movements with her head as though gathering litter around herself as the hen is apt to do in a nest-box. This behaviour often changes into dust-bathing [263]. A sample of birds watched showed extreme conservatism in their choice of the part of the cage where they laid their eggs, though there was some variation in the stance adopted for oviposition. Generally the egg was dropped from heights of 25 to 75 mm, but greater heights were recorded. Under these conditions the birds made no attempt to sit after laying, although in nest boxes most birds sit for appreciable periods after laying. A striking feature of many of these birds before oviposition was their intense stereotyped behaviour. This needs further investigation, not only in relation to energy loss but possibly also in relation to animal welfare and ethology [263].

Parental behaviour

The endocrine basis of both incubation and brooding has been well reviewed by Eisner [59]. She points out that incubation is much more difficult to induce than brooding. Hens easily become broody as a result of exposure to chicks, but respond much less readily to accumulations of eggs. Prolactin can be effective. Cocks and capons can be induced to take care of chicks by prolactin, but not to incubate eggs. Goodale [77] cites the case of a cock that hatched a clutch of eggs, but this appears to be the only case published.

There is genetically determined variation in parental behaviour. Hens of certain genotypes incubate eggs after receiving prolactin alone only if they are in active egg production at the time [59]. Non-laying hens in moult can be induced to incubate if prolactin

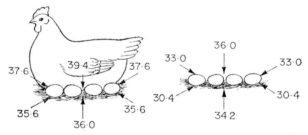

Figure 47. The temperatures in the hen's nest during incubation and
while the hen is absent.
(From Tretyakov [228].)

treatment is combined with cooping in darkness and warmth,
although such cooping alone is ineffective.

Hellwald [109] recorded the food intake of three incubating
hens. They generally took only one-fifth of their normal intake
and took more water than solid food. On some days they took no
food at all. They lost from four to twenty per cent of their body
weight during this period.

Kuiper and Ubbels [137] investigated the temperatures at which
hens incubated their eggs. The maximum temperature was 39·3°C,
and this occurred only where the egg was in contact with the hen.
Tretyakov [228] also investigated the temperature of eggs in natural
conditions of incubation (figure 47), the eggs were turned every
hour. In cold weather and during the early days of incubation the
birds sit more.

Burrows and Byerly [30] induced broodiness in Rhode Island
Red hens by placing them singly in a dim coop with baby chicks.
The dimness eliminated other visual stimuli. When the coops were
replaced by darkened boxes with high temperature and humidity,
broodiness was induced in two White Leghorn hens—a breed that
is classified as non-broody. Furthermore broodiness was induced
in a twelve-week-old cross-bred pullet by these conditions. Ramsay
[182] induced broodiness in Cochin Bantam hens by keeping them
with small chicks but not excluding other visual stimuli. There was
great variation in the time taken by the hens to become broody.
He divided the whole behaviour pattern into four stages: brooding,

tidbitting, clucking, and normal brooding. By 'brooding' he meant the assumption by the hen of the brooding posture, in which the wings are typically spread and lowered to invite brooding. At night the hen stands instead of squatting and the feathers become ruffled. In the tidbitting stage the hen gives a special food call, picking up food and dropping it. During this stage the hen will come to the assistance of a chick that has been picked up and is giving distress calls. The clucking stage and the fourth stage according to Ramsay's classification grade into one another, the clucking becomes loud and continuous in the fourth stage. At least one hen clucked before accepting a chick, so that the order of appearance of Ramsay's stages may be regarded as only provisional, since few birds were used.

Hen-chick relationship

In the studies on imprinting much work has been done on the analysis of the stimuli that initiate the approach of the following response. James [121], working on the hypothesis that a flickering light should have the same effect on the retina of a newly-hatched chick as a moving object, found that two-day-old chicks approached intermittent light showing through four small holes at the end of the runway. Smith [210], who also found that young chicks approached a flickering patch of light, reported that a white disc with a forty-five-degree black sector, placed at one end of a 3-m runway, caused chicks to approach when the disc was slowly rotated so that the black sector moved round and round. The flickering light and the rotating black sector were equally effective. Later, Smith and Hoyes [211] found that chicks approached other patterns moving at right angles to the chick's line of vision.

In chapter 4 the analysis by Collias and Joos [43] of the calls that attract young chicks was described. Sluckin [209] has given evidence for the interdependence of auditory and visual stimuli. Sometimes auditory stimuli are more important than visual clues in eliciting following after imprinting has taken place. Twenty-four-hour-old chicks imprinted first to a red cube emitting broody calls, when retested twelve hours later showed greater response to an orange cone emitting broody sounds than to the red cube either emitting duck's twitters or without any sound [66].

Chicks recognize their dams by various means. Collias [41] put chicks of different broods together in the dark, there was a tendency for the chicks to go to their own hens. But visual stimuli are also important for individual recognition: Collias also took chicks from hens of three colours and released them in a pen containing three hens of the same colours as the mothers. The chicks from a black hen went to the black one, those with a red mother went to the red hen, and those from the white to the white. Some chicks made mistakes.

The question whether a hen recognizes her chicks is difficult to answer, for, as Bruckner [29] points out, the structural and vocal traits of chicks are continually changing. He suggests that a hen knows her brood by a 'complex recognition-impression' of the whole brood. If a strange chick has the appropriate set of characteristics it is accepted, but generally a hen treads on, or pecks at, a strange chick. To be accepted a strange chick must have the same behaviour as her own chicks, its colour is in this respect less important. If it is more advanced or retarded than the foster brood it is unlikely to be accepted. After observing the positions taken by the chicks in the nest Bruckner concluded that a hen has no favourites within the brood, and it is a matter of 'first come, first served'.

For the first ten to twelve days the chicks are in close contact with the hen. Then they enter a dispersal stage [29] in which they feed independently of the hen but still sleep and warm themselves under her. This stage lasts until the chicks are six to eight weeks old. The time of dissolution of the brood varies. Bruckner reports several broods which endured for twelve to sixteen weeks, but usually the hen ceases to be broody before then and drives the chicks away.

5: Maintenance Behaviour

Feeding behaviour and its development

The feeding behaviour of the Burmese Red Junglefowl has been fully described [136]. No comparable description exists for the domestic fowl but, as the dissimilarities are few, Kruijt's description can form the basis of a description of feeding behaviour in the domestic species. Apart from pecking and swallowing the Junglefowl has two main behaviour patterns: head-shaking and bill-beating, that serve directly to break up the food. Another group cleans the bill and head: this includes head-scratching, bill-wiping, and bill-scratching. Ground-scratching, often associated with feeding, can make the food more accessible. The Junglefowl three-day-old chick performs it in full, evidently without any special releasing stimulus; Kruijt suggests that it is automatically coupled with the performance of pecking movements. Ground-scratching by the domestic chicken too seems to be independent of special stimulation.

Food-running is a very striking feature of feeding behaviour; the bird runs around with a large food object or living prey in its bill, peeping continuously. It is then usually chased by other chicks which grab at the food whenever possible. Spalding [214] described this behaviour in the domestic fowl and reported that it occurred in chicks kept in isolation. Baeumer [15] states that it is released by the taste of meat only, but Kruijt's observations and incidental observations by the author indicate that it is released by a large number of stimuli. Kruijt suggests that its adaptive value lies in the high probability that the chase will lead to tussles over the food object which will then be torn into edible portions.

Many experiments have been made on the development of pecking by domestic chicks. Spalding [214] covered the heads of young

91

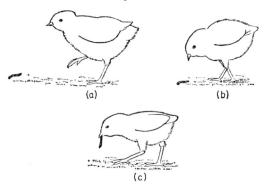

Figure **48**. Behaviour associated with food-running. (a) Cautious approach in alert posture with monocular fixation; (b) posture while giving trill call; (c) running with meal-worm in bill.
(From Kruijt [136].)

chicks with hoods during their first one to three days; when the birds were released on to a sheet of white paper covered with insects, both dead and alive, they pecked fairly accurately at the insects within fifteen minutes. Swallowing, he reported, was not so easily achieved.

In much of the work on the development of pecking, a technique widely used, to determine the amount of learning involved, has been to keep chicks in the dark from hatching and to test their pecking responses at various ages in the light [47, 171 and reviewed in 249], and to compare their accuracy with that of control chicks, either kept in the light with unlimited access to food, or given limited amounts of practice at feeding and pecking.

A general criticism of the technique is that rearing in darkness may affect the normal development of the eye. Hess [110], however, designed an experiment that overcame this difficulty. He reared his chicks with fitted prismatic lenses which displaced the visual image without interfering with the normal development of the eye. One group had the visual field displaced to the right, one to the left, and a third served as a control.

After a six-hour period of adjustment the chicks were tested at about twenty-four hours of age and again at three to four days. At the second test the scatter of pecks was much reduced in all three groups, but the pecks of the experimental birds were all displaced in the direction dictated by their lenses. On the other hand some earlier work [47 and 171] had suggested that practice does make a contribution and this has been confirmed by a recent experiment by Rossi [191]. His chicks from hatching wore hoods containing prisms that caused a lateral optical displacement of 8·5 degrees and after eight days showed significant adaptation. Furthermore after substitution of these hoods for ones with clear blank plates instead of prisms the chicks showed over-compensation in their pecking according to their previous experience.

Katz and Keller [128] have investigated the aiming and pecking ability of adult fowls with a revolving target with a restricted field of vision. A horizontally revolving stage with grain scattered over it could be moved at controlled speeds. With the stage stationary a hen pecked at the rate of 3·5 times per second, and this pecking rate did not alter when the stage was moved at 13 cm per second. When the stage movement was 33 cm per second it pecked at 1·8 and 2·0 times per second. It stopped entirely when the stage was speeded up to 66 cm per second. At medium speeds the bird followed the grain with small jerky movements and the authors suggest that optical orientation follows only between single jerks. When the field of vision was decreased to 22·5 cm × 15 cm the pecking efficiency decreased. When the grain appeared on the bird's right and disappeared on its left, and moved at medium speeds, the bird pecked once per second. When it had to face the oncoming grain it pecked nine times per second, but when the grain moved away from it, the bird refused to peck at all. Hutchinson and Taylor [114, 115, 116] also investigated the accuracy of pecking by adult birds with a horizontally revolving stage and a restricted field of vision. They found individual differences between birds as well as differences between the sexes: hens were more accurate than cockerels. With this type of technique a negative correlation was found between pecking accuracy and comb size, and trimming of the bill impaired the accuracy of most of the birds temporarily. A slow motion film of these birds pecking at wheat grains on a board showed that striking occurs at a distance of 1 to 3 cm. The strike, which has considerable force is performed while the nictitating

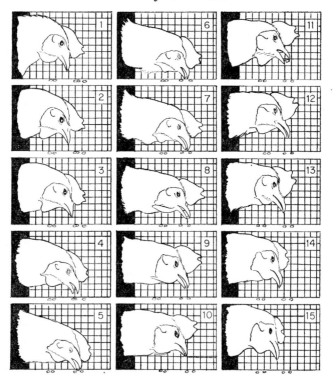

Figure 49. A film showing a hen pecking at grain. Notice the nictitating membrane is drawn over the eye in frames 4 to 7.
(From Hutchinson and Taylor [116].)

membrane is closed. The grain is grasped about the short axis by moving the upper mandible to the lower, or the lower to the upper, or by moving both together. After prehension the head is withdrawn upwards and backwards, and usually several jerking movements (deglutitions) are necessary to guide the grain to the throat. In this phase the sensory control appears to be poor, for even if the grain is missed the bird may go through one or two deglutitions.

These authors also tested their birds' pecking responses at high temperatures. Pecking accuracy was usually higher when rectal temperatures were up to 42 to 43°C than when they were normal.

At these high temperatures birds usually pant freely, the authors suggest that the panting 'centre' [238] is inhibited by feeding.

The motivation of feeding

Internal processes

Our knowledge of the internal control of the feeding behaviour of fowls is at present not nearly as detailed as it is for the rat. Lepkovsky and his co-workers have been investigating the role of the hypothalamus, and find that the organization of feeding behaviour is very different in the two species. Lepkovsky and Yasuda [144] report that the differences between the rat hypothalamus and fowl hypothalamus are too great for useful comparisons. They were

Figure 50. Diagram of bird (left) with lesion near Median eminence showing the effects on secondary sexual characters compared with normal bird of same sex and breed on the right.
(From Lepkovsky and Yasuda [144].)

able to produce hyperphagia by means of lesions near the median eminence. About half the hyperphagic birds, however, showed signs of gonadotrophic imbalance, described by the authors as caponization. However, these lesions may have produced hyperphagia indirectly through hormonal action. For example, exogenous oestrogens lead to increased food intake [151]. Furthermore van Tienhoven and Cole [232] found that a line of obese pullets with small combs had thyroid malformations. The authors attributed them to decreased TSH secretion. The median eminence is connected with both gonadotrophic and TSH secretion, so that the hyperphagia and obesity could have been due to the indirect effects of these hormones.

Feldman and his co-workers [64] made bilateral lesions in the hypothalamus of each of a large number of fowls, and produced eighteen aphagic birds. The lesions did not necessarily involve the lateral hypothalamic area, the destruction of which in the rat causes aphagia, but some of the aphagic birds had lesions in the anterior or posterior regions of the hypothalamus. Von Holst and von St

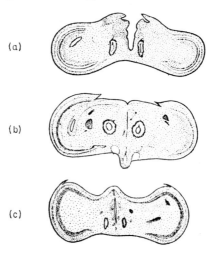

(a)

(b)

(c)

Figure 51. Sections of hypothalamic areas showing bilateral symmetrical lesions causing aphagia in fowls. (a) Lesions in the posterior hypothalamus just posterior to the hypophysial stalk. (b) Lesions in posterior hypothalamus. The specimen has been cut obliquely from the dorsal surface so that the posterior commissure and the pituitary stalk are both present in this section. (c) Lesions at the level of the posterior commissure.
(From Feldman, Larsson, Dimick, Lepkovsky [64].)

Paul [236] elicited elements of feeding behaviour by stimulation of parts of the brainstem in free moving birds; but they give no anatomical details.

The sensory feed-back mechanisms of the rat [164] make an interesting contrast with that of the fowl and pigeon. Fisher and Weiss [67] found that the crop of the young chicken does not control food intake over a long period, but it may control food intake in the short-term. This is suggested by two studies, one on pigeons. The crop of a hungry pigeon undergoes contractions which are correlated with behavioural restlessness, particularly in decorticate birds which are less attentive to external stimuli [189]. Water in the crop of these birds leads to crop relaxation and cessation of restlessness. Furthermore, there appears to be a quantitative relationship between the amount of water in the crop and relaxation of the crop. Solid food, however, is a more effective relaxing agent than water. Rogers [189] reported that the stomach and gizzard of the hungry pigeon show vigorous contractions and that these are accompanied by behavioural restlessness.

In the second study young chicks were kept in activity cages and, when deprived of food, increased their activity during four days' deprivation [31]. However, as with the rat [164] the amount of food subsequently taken is not necessarily correlated with the length of the fast. In the rat any correlation disappears after fasts of more than six hours. Wood-Gush and Gower [262] found a similar situation when adult cockerels were given food as mash: consumption during thirty minutes was greater after fasts of twenty-four or forty-eight than after two hours' deprivation, but there was no difference between the amount consumed during the half hour feeding period after the two long fasts. Neither were there any differences in the amounts consumed over the first three minutes of the meal by the birds after these long periods of deprivation. Furthermore, even when the birds' intake was measured over periods up to seventy-two hours, there was no difference between the birds fasted for twenty-four hours and those fasted for forty-eight hours. When given pelleted food, on the other hand, the birds fasted for forty-eight hours ate more in the half-hour meal than those fasted for twenty-four hours. The intake of pellets, as was expected, was far more rapid than the intake of mash, and the crops of the birds after feeding on pellets for the half-hour were fully distended, while after the mash the crops were far from full. The crop may therefore play a part in governing the size of the first meal, and the birds may have tolerated greater distension after forty-eight hours of deprivation than after twenty-four hours.

The finding that the birds fasted for forty-eight hours did not

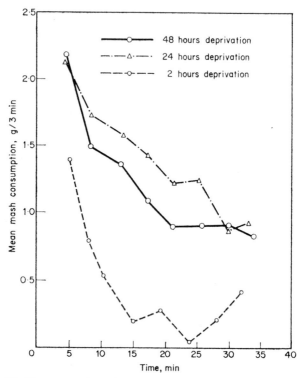

Figure 52. The rate of mash consumption by cockere ls over a 31 minute period after 2, 24, and 48 hours of food deprivation. (From Wood-Gush and Gower [262].)

make up their deficiency in seventy-two hours suggests that there are two mechanisms controlling food intake, one short-term and the other a long-term one governing body weight. A similar dual control of the feeding of rats has been proposed [221].

Measures of feeding behaviour, other than amount of food consumed, include the rate of pecking, the strength of the peck and the rate of mash consumption [262]. The last two are not consistently related to the length of fasting, whereas the rate of pecking mash follows the amount of mashed consumed: pecking rates after two hours' deprivation were less than after fasts of twenty-four or forty-eight hours. Again, pecking rates were the same after twenty-four and forty-eight hours of fasting.

The regulation of feeding by fowls is at present very poorly understood. In comparison to the rat a great deal more is, however, known about food conversion, and the fowl therefore offers some advantages as an experimental subject in this respect. The rate of food conversion in relation to the growth and completion of the ovum has been investigated by a number of workers [231]. The rate of incorporation of elements sometimes is surprisingly fast and suggests that a study of the relationship between feeding behaviour and egg formation might prove interesting.

External factors

Feeding behaviour is influenced by many factors other than internal state. For a social animal like the fowl social factors can be extremely important. The appetite of a bird can be increased by taking away the food at intervals or by introducing another hungry bird [27]. Tolman and Wilson [226] deprived chicks of food for 0, 6, 12, or 24 hours and then examined their food intake over one hour under various social conditions in which a chick either had 0, 1, 4 or 16 companions. The presence of a single companion was accompanied by increased food intake only in the six-hour food-deprived group. An increase in the number of companions to sixteen did not result in greater social facilitation in any group. The sight of another chick through glass did not increase food intake: for social facilitation to be effective contact is evidently necessary. The state of hunger of the companion too is very important.

It was mentioned earlier that the young chick pecks wherever the hen pecks [41] and in this way learns what to eat. Tolman [227] investigated the effects of tapping sounds on pecking by young chicks. He had earlier found that the sight of a model hen pecking at the ground induced chicks to peck. In the absence of the model hen tapping sounds at regular frequencies at first increased pecking by paired chicks, but habituation occurred within twenty minutes. The optimum frequency was between 60 and 120 taps per minute. Tolman concludes that the visual stimulus of the feeding companion

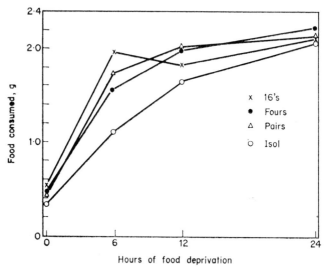

Figure 53. The amount of food consumed as a function of hours of deprivation and social conditions. Chicks tested under isolation, or in groups of 2, 4, and 16.
(Adapted from Tolman and Wilson [226].)

is the main one to elicit feeding and that the auditory stimulus enhances its effect.

Many behaviour patterns have a circadian rhythm and external stimuli are often the proximal or initial pace makers [12, 13]. The feeding of fowls was studied in natural light for two consecutive days in midwinter and another two such days in midsummer in Scotland [255]. Feeding at the mash hoppers declined in the afternoon in the summer but in the short winter days it was fairly constant. Feeding and other activities recorded in this study were, however, not completely tied to the light-dark cycle. Roosting occurred in summer while it was still early evening, and the birds were active on the floor of the pen before it became light in midwinter.

The physical properties of the food stimulus can affect feeding behaviour. Bayer [22] suggested that the greater the pile of food set before the hen, the more she eats. Ross and his co-workers [190] investigated this hypothesis with four-week-old chicks that had been deprived of food for four hours. The amount eaten by a chick was found to be linearly related to the amount presented, but it may have been easier for the chicks to eat from the larger piles. Nevertheless the question of visual stimulation in relation to feeding requires more investigation. Intimately connected with it is the study of food preferences which is discussed below.

Food preferences

Engelman [60] found that his birds had a strict preferential order for cereal grains as follows: wheat > maize > rye > barley > oats. When these cereals were presented to the same fowls in powdered form, wheat, rye, oats, and barley were eaten at the same frequencies. Ground maize was less popular. Engelmann also carried out experiments with artificially made grains. When 'wheat' and 'rye' grains were made from rye meal, the 'wheat' grains were significantly more popular. When 'rye' grains made from wheat meal were presented at the same time as 'rye' grains derived from rye meal, they were equally popular. 'Wheat' grains made from rye meal were more popular than 'rye' grains from wheat meal. Finally, when 'wheat' grains from rye meal and real rye grains were offered, the 'wheat' grains were preferred. Engelmann also dyed wheat, maize, and rye grains in eosin. At first the fowls refused them, but after a day's hunger they accepted them. Not all the colours were identical and no check on brightness was made. After the fowls had learnt to accept them the order of popularity remained. When the ground forms of the cereals were dyed and artificially made to look alike, they were all accepted at equal frequencies. Engelmann concluded that the birds formed preferences on the basis of form and colour with the possible aid of tactile impressions but not of taste.

More recently, however, it has been shown that the fowl has a sense of taste (chapter 2). Kare and his co-workers [124] tested young chicks for their preferences for thirty-two flavours. Each flavour was in solution and its acceptibility was compared with that of distilled water. The positions of the water troughs were changed every three or four hours and losses due to evaporation were taken into consideration. The oils of costus, patchouly, and sandalwood were all rejected. In some respects the preferences were very different from those shown by man: sucrose was slightly preferred but saccharine was rejected to a moderate degree. Later Kare and Medway [125] compared the chicks' preferences for sugars at different concentrations. Dextrose in concentrations up to twenty-five per cent was neither preferred nor rejected, compared to water. Sucrose was similarly accepted up to concentrations of twenty per cent but mildly rejected at a concentration of twenty-five per cent. Xylose was progressively rejected at concentrations of five per cent or more. Maltose was generally preferred at concentrations up to $12 \cdot 5$ per cent. Lactose, galactose, raffinose, and fructose were neither preferred nor rejected at concentrations of five per cent while arabinose and mannose were mildly rejected at these levels.

On the mechanism of selection, Kare and Medway suggested that the following features should be considered: osmotic pressure, viscosity, nutritive value, tactile stimulation, differences in the refractive index of the solutions, differential effects on the motility of the intestinal wall and villi. They concluded, however, that the preferences and rejections of these sugars could not be related to any of these factors.

Fuerst and Kare [73] tested the effect of pH on the fluid tolerances and preferences of chicks over the first eighteen days of life. The chicks showed a wide tolerance of pH: their acceptance ranged from pH_2 to pH_{10}. From pH_2 to pH_1 their acceptance fell sharply but from pH_{10} it fell gradually to pH_{13}. There were also changes with age: at first the chicks took more of the alkaline solutions but after six days of age their preference switched to the more acid solutions.

The ontogeny of the food preferences of the fowl has been little studied, but some work has been done on the colour and shape preferences of young chicks. Hess [112] found that chicks peck at some colours more than others, with one peak of response occurring in the orange region of the spectrum and a second in the blue region. This finding is surprising in view of the report that adult birds do not readily peck at blue objects [239].

This preference for certain colours resembles the form preferences [62] mentioned in chapter 2 in being, evidently, independent of previous experience of the colours. By contrast, it is possible to change the early food preferences of chicks by coupling the presentation of preferred foods with a noxious stimulus consisting of an injection of ten per cent NaCl directly into the crop [32].

Specific hungers

The fowl, under quasi-commercial conditions, has been reported to be able to select a balanced diet [74, 78, 79]. However, the experiments on which this claim is based have certain defects: the food was in the form of plant or animal material and an adequate, or nearly adequate, diet could have been obtained by a random choice, the bird's preferences could have fortuitously led them to a correct diet, no detrimental or non-nutritive substances were included. In an experiment by Funk no salt was taken by the birds for two weeks [74]. Graham [79] reported that chicks hatched in February took more cod liver oil than others hatched in May. This work needs to be repeated, for this finding could be explained on the basis that cod liver oil is less palatable in hot weather.

Jukes [123] allowed chicks deficient in one vitamin a choice between the deficient diet and the same diet reinforced with the vitamin. He found no preference for vitamin A or lactoflavin. Kare and Scott [126] offered groups of sixteen-day-old chicks a choice between a standard chick diet and one deficient in protein or cereal. Seven alternative diets were used, and the chicks' preferences corresponded in no case with the nutritional need. Similarly from experiments in semi-intensive agricultural conditions, in which the hens were allowed a limited choice, the birds did not take the best possible diets for their needs [153].

A good indication that the fowl can adjust feeding to a special need is given in an early paper by Hellwald [109]. He deprived eight hens of calcium for nine days. All birds were then fed macaroni but in the macaroni of one group small bits of egg shell were hidden. In the afternoon after the feed, both groups were placed in front of piles of egg shells. The group that had received none in the morning took 91 g while the other took 27 g. When the experiment was repeated a week later similar results were obtained.

Wood-Gush and Kare [261] confirmed that the fowl can adapt its calcium intake to need. Growing broilers were deprived of calcium until their blood-calcium levels were lower than those of control birds. The birds were then allowed a choice between two mashes differing only in the absence or presence of either one per cent or four per cent calcium carbonate. Each bird was individually

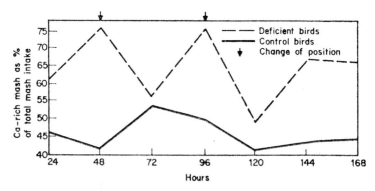

Figure 54. The consumption of calcium-enriched mash expressed as a percentage of total food intake by calcium-deprived and control fowls (under a two choice procedure); the positions of the troughs being exchanged every forty-eight hours.
(From Wood-Gush and Kare [261].)

caged and allowed a choice between the calcium-enriched mash and the deficient mash presented in identical troughs. The positions of the troughs were changed every twenty-four hours in one experiment and every forty-eight in another. In both experiments the calcium-deficient birds took significantly more of the calcium-enriched mash than the controls. In a third experiment calcium was offered as calcium lactate solution. All the broilers rejected it almost entirely. The calcium-deficient birds explored potential food objects in a strange environment more than control chicks did. However, the objects explored most did not necessarily resemble any well-known natural source of calcium. Probably the deficiency leads the animal to explore more and subsequently to learn which diet is beneficial.

Drinking

Spalding [214] thought that the newly hatched chick had to learn to drink. Lloyd Morgan [149] observed that chicks never peck at a sheet of water even if they are thirsty and are standing in it, but that they will peck at any material or at bubbles in the water. As soon as the bill is wet the chick begins to drink. Kruijt [136] suggests that this is also the case in the Junglefowl chick.

A recent study by Rheingold and Hess [186], however, suggests that drinking by inexperienced chicks may be elicited by a visual stimulus. They set out to discover the visual properties of water which attract chicks. They assumed that any of the following properties of water, either alone or collectively, could attract the chick: colourlessness, transparency, reflecting surface, or movement. Six stimuli were chosen as possessing some of these properties. One hundred three-day-old chicks were used. The experimental birds were deprived of water until after their first test. They were again tested at seven days of age. Twenty-eight control chicks which had had experience of food and water were also used. Each bird was tested individually for its initial preference from the following groups of stimuli; water, blue water, red water, polished aluminium, mercury, and plastic. Their order of choice at three days of age was mercury, plastic, blue water, water, metal, and red water. Three times as many chicks chose mercury in preference to water and more than twice as many chose plastic. At seven days the order was only slightly changed: blue water superseded plastic, and red water supplanted metal. The control chicks had the same preferences. The initial responses to the chosen stimuli were reported to be true drinking responses: the movement of the head was slower than in pecking and the beak was pushed forward and upward.

Little is known about the internal control of drinking by fowls. Adipsia has been produced in one fowl by bilateral lesions in the lateral area of the dorsal hypothalamus: although the bird pecked at water it did not drink. It was kept alive for several months with water given through a tube, and was able to eat normal mash. Water enriched with glucose or milk was rejected [145].

Two studies on the circadian rhythm of drinking have been

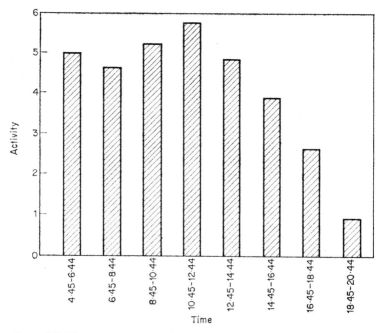

Figure 55. The visits per cockerels when aged 14 to 15 weeks to the water fountain at different times during the day as measured by photoelectric systems in a pen. The temperatures averaged from 31·1 – 35°C during the period.
(Adapted from Siegal and Guhl [203].)

made. Siegel and Guhl [203] recorded the activity about a drinking fountain of nine- to fifteen-week-old cockerels during the summer, when the temperatures varied from 28·9 to 35°C. Activity was greatest from 10·45 to 12·45 hr. The other study is the time-lapse film referred to earlier [255]. The frequency of drinking varied with time of day; peaks were related to feeding. Two of the birds could be identified in the film, and they revealed considerable differences in drinking frequencies from the group average during the two days of filming.

While the film revealed considerable competition among birds at the food hopper, drinking appeared to be free of it; the space allowances in both places were very generous.

Comfort and grooming behaviour

Few of the movements that domestic fowls make to groom the body and care for it have been analysed in detail. Kruijt's description and analysis of the corresponding movements of Junglefowls is the most detailed account for gallinaceous birds that we possess and, while the domestic fowl probably differs in minor ways, Kruijt's description is used in this section.

Under comfort movements Kruijt lists stretching, yawning, preening, dustbathing, and shaking. He distinguishes two types of stretching by the Junglefowl: bilateral and unilateral. It is very uncertain whether bilateral stretching, in which the wings are half opened and stretched upward and forward, even occurs in the domestic fowl. A very similar movement does occur, but it appears to be confined to conflict situations and is more likely to be an incomplete form of wing-flapping (see chapter 3). He describes preening as consisting of six main movements. Yawning is certainly seen in the domestic fowl and may occur in conflict situations.

The frequency of preening by the domestic fowl was estimated in Wood-Gush's study, by time-lapse photography, of a small group of adult birds in a pen. The total time spent preening by the group varied from 1·8 per cent to 4·0 per cent of the twenty-four hours, and one of the identifiable hens spent 5·7 per cent of the twenty-four hours preening. The birds were free of ecto-parasites and the frequency of preening varied with the time of day, and with the frequency of sleeping. Evidently, as the bird becomes sleepy, many stimuli in the immediate environment lose effectiveness while tactile stimuli gain in relative strength and lead to preening. If the bird is waking gradually these stimuli are similarly among the first to become operative.

Dustbathing is a highly social event in the fowl in that a dust-bathing bird is usually joined by others, but birds in battery cages may perform it. In the Junglefowl Kruijt distinguishes the following component movements: bill-raking, vertical wing-shaking, scratching, lying on the side, and head rubbing.

Feather ruffling may be another comfort movement. Others listed by Kruijt are head-shaking and tail-shaking. They appear in conflict situations and head-shaking by cockerels may occur in response to the crowing of other males. Their appearance as comfort movements is probably rare.

Sleeping has not been widely investigated physiologically in the fowl, but several studies on EEG patterns of sleeping and awake birds have been done [167]. Ookawa and Takenaka [170] found different EEG patterns from surface recordings and from the ektostriatum and ractus fronto archistriaticus. However, since sleep or drowsiness may also be caused by conflict or thwarting it is imperative that workers in this field give more details about the environment of their birds before definitive conclusions can be drawn.

The frequency of sleeping and the depth of sleep as gauged by gross body movements were investigated in the time lapse film. The birds tended to go to roost while it was still light even on the eight-hour winter days. In summer they did not leave the roosts until sometime after dawn and they roosted while it was still very light. Hence sleep is partly independent of light. Palmgren [172] studied the sleep of wild birds under laboratory conditions and concluded that it occurs as an autonomous periodic function, although largely controlled by light. Aschoff and Von Holst [11] studied the flight times of Jackdaws (*Corvus monedula*) to and from the roosting places in Heidelberg for two years. These activities depended not only on endogeneous periodicity but also on synchronized external factors, one of which was the light intensity which varies with the season. Wilson and Woodward [242] kept hens in darkness for five weeks. Although the attendants used guarded flash lights for their work, the light of the room, as registered by a film exposed for one hour, was about 0·0002 foot-candles. Egg production declined to an average of twenty-three per cent and in the last week rose to forty-three per cent for one batch of birds. Most of the birds lost weight, and the authors gained the impression that although the birds called they were less active than birds kept in the light. Light perhaps plays only a subsidiary role in the activity rhythm of fowls.

6: Learning

The involvement of learning in the behaviour patterns of the fowl is discussed in the other chapters. Here we shall discuss experiments directed exclusively to the study of learning with the fowl as the experimental animal. The terms used here follow the definitions given by Thorpe [222].

Habituation

Although this type of learning must have played a very important part in the domestication of the fowl it has not been much studied. A chick ceases to respond to certain types of acoustic stimuli far more quickly than a rat in very similar conditions [218]. Although no control experiments were done to test for sensory fatigue the rapidity with which the chicks ceased to respond (one trial) suggests that the habituation was due to a central, not a peripheral change. In another experiment chicks of White Leghorn, White Leghorn–New Hampshire cross and Bantam stocks, aged 50 to 340 hours, displayed a decline in avoidance responses to a standard stimulus presented at intervals of not less than five hours [196]. Again, sensory fatigue can hardly have been responsible. Another example comes from a study by Gilman, Marcuse, and Moore [76] concerned primarily with the relationship between hypnosis, fear, and perception. In one experiment five birds were hypnotized by the same experimenter in the same locality by being placed in the identical dorsal position on each trial. At the first trial, four out of five birds were susceptible, but after thirteen days and twenty-five trials none was susceptible. At the twenty-sixth trial the experimenter, locality, and body position were all changed and the incidence of susceptibility returned to four out of five. At the next trial, the experimental conditions were again identical to those used in the first twenty-five trials and the incidence of susceptibility fell to zero again. A further trial similar to the twenty-sixth followed and all five birds were found to be susceptible. The authors state that the change of experimenter was the main factor, but some body positions were more efficacious in making the birds susceptible to hypnosis and so the experiment needs elaboration. In another experiment fourteen birds were hypnotized in forty trials over a twenty-one-day period. Several body postures were used, but for any one bird the same position was employed. The incidence of susceptibility fell from over seventy per cent on day one to ten per cent on day twenty-one. A further eight birds were given thirty trials during one day and the percentage of susceptibility was almost the same in

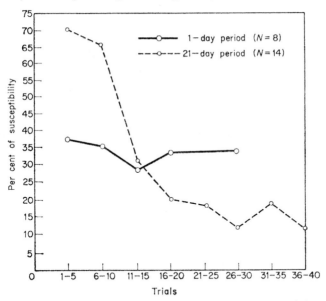

Figure 56. The effect of repetition on susceptibility to immobilization when 40 trials are given over a 21-day period, and when 30 trials are given over the course of one day. N = the number of birds in each experiment.
(From Gilman, Marcuse, and Moore [76].)

the first five trials as in the last five. Evidently, the decline in susceptibility during the twenty-one days was not due to sensory fatigue, this, if important, would have shown up in the birds tested thirty times in one day.

Conditioning and trial-and-error

Examples of classical conditioning are probably fairly common in typical fowl houses, for example, the sight of food can elicit food calls and certain sounds such as metal buckets being filled with food in another room may regularly precede feeding and in time come to elicit food calling. In laboratory conditions several experiments have been carried out. Watson [241] trained six chicks to change their respiration rate; the unconditional stimulus was an electric shock and the conditional stimulus the sounding of a bell. A tapping sound attracts young chicks and Grindley [81] using this as the unconditional stimulus for the approach response of young chicks, trained them to approach a hidden food source, and by coupling the sound with the sound of a horn, trained them to approach the hidden food source when the horn sounded. Zawadovsky and Rochlina [265] trained fowls to discriminate between one of two windows of a cage. They were rewarded with food if they approached one window when the stimulus was 100 beats of a metronome. Once they had learnt this, however, it was difficult to train them to put their heads through the other window with forty metronome beats per minute as the stimulus. When one cock had acquired the first response it was then successfully trained to approach the second window at the sound of a bell.

Examples of operant conditioning involving the use of a Skinner box are rare in the fowl. Lane [142] carried out a very interesting experiment in which he imposed particular patterns on the rate of chirping of two Bantam chicks. Their rate of chirping was first determined in a free-feeding situation and in the same situation with food absent, and found to vary between twenty-four and twenty-seven chirps per minute. When placed on a fixed ratio schedule of four seconds of feeding after every twenty chirps the chirping rate rose to 115 chirps per minute, this change in the rate of chirping was not due merely to a change in the situation for, when an empty tray was presented instead of food, the rate of chirping fell to eight per minute. When neither food nor tray were presented, chirping fell to zero within thirty minutes.

In an interesting experiment on operant conditioning [8] chicks aged four days had electrodes implanted in the medial forebrain bundle. Flexible leads suspended from above allowed the chick free movement and by pecking at one of two discs it could stimulate itself electrically; the discs gave currents differing in duration or intensity. Testing was done when the birds were between five and fourteen days old. Control chicks had similar implantations and were likewise connected to overhead leads, but they received no

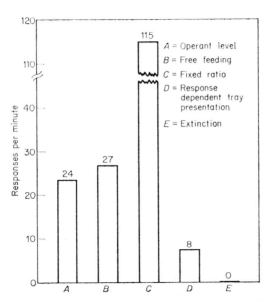

Figure 57. Rates of 'chirping' obtained from one Bantom chick under experimental conditions in which the chick was rewarded with food for chirping (column C) compared to chirping rates with food freely available (column B), or no food available before training (column A). When no food was presented after training chirping fell (columns D and E).
(From Lane [142].)

current on pecking a disc. Each chick also served as its own control in that its rate of pecking without stimulation was recorded. The pecking rates of the six experimental birds were higher than those of the controls and also higher than their own unrewarded rates. The rates on stimulation reached 260 pecks/10 min and so were greater than the rates (150/10 minutes) of rats with electrodes in the medial forebrain bundle.

Discrimination learning has been reported by a number of workers [5, 37, 80, 168, 183, 217, 266]. In some experiments both reward and punishment have been used. Cole [37], for example, rewarded his chicks for a correct discrimination by allowing them to escape from an electric shock while a wrong discrimination failed to lead to escape. Generally the stronger the current, the quicker the discrimination was learnt, but some chicks never learnt under these conditions. Grindley [82] attempted to investigate the relationship between rate of learning and the amount of positive reinforcement in the form of food. His work was considerably refined by Wolfe and Kaplon [245] who attempted to distinguish between the amount of food eaten and the amount of consummatory activity. Essentially, their experiment consisted of testing three groups of chicks in three learning situations, each group received a different reward. Group 1 received one large piece of popcorn per trial, group 2 a piece of popcorn, a quarter the size of that received by Group 1, while Group 3 received four small pieces which equalled the amount received by Group 1. All the birds were deprived of food for three hours before the experiment, which was done three times with twenty-four chicks aged forty-one, fifty-seven, and seventy-three days.

The problems consisted of learning to run down an alley to get the food, a simple detour and a T-maze. Each problem was given to each bird seven times. Out of the twenty-one tests Group 3 was more efficient than Group 1 in six tests, and in all the others it tended to be the more efficient. Group 1 and 2 did not differ. Unfortunately the authors did not compare Groups 2 and 3 statistically, but in twenty out of the twenty-one cases the mean scores of Group 3 were superior to those of Group 2. Within the limits of the amount of food given and the degree of food deprivation, the amount of consummatory activity was evidently an important element; but the possibility that perceptual factors influenced the chicks must not be neglected.

Nash, Warren, and Schein [168] reported that White Leghorn hens retained discrimination learning nineteen days after the problem had been learnt even when eighteen other discrimination problems had been mastered in the interval. The effect of incubation at 41°C for part of the incubation period has been found to affect the learning and retention of an alternation problem [101]. Ten chicks were incubated at the optimum temperature (37·5°C); another ten at 41°C for their first twenty-four hours, and a third group of similar size incubated for their first forty-eight hours at 41°C. The controls learnt significantly better than the experimental groups, the second of which failed to attain the criterion of learning. Three control birds and three from the one-day high temperature group were tested for retention of the problem two weeks after learning. All the controls showed better retention than the one day group which, however, showed some retention. Since incubation at this temperature also affects growth adversely [102], the detrimental effects on learning and retention are not necessarily attributable to changes in the CNS, for other peripheral changes could have affected learning ability. The retention of an elementary detour problem was found in Rhode Island Red chicks six days after the problem had been learnt to the criterion of ten successful successive runs, when the chicks were between four and nine days [200].

The formation of learning sets has been studied in both adult fowls and chicks [5], and two-week-old chicks were as good as adult hens. White Plymouth Rock fowls aged four months have also been tested for their ability to form learning sets; Plotnik and Tallarico [178] equate the birds' ability in this respect with the raccoon, cat, and marmoset. Fowls have also been tested for reversal learning [14 and 240].

The ability of the fowl to learn to peck at every second grain in a row has been shown by Honigman [113]. In the training situation the alternate grains were glued down, but care was taken to eliminate visual clues which would help to distinguish the loose grains. Great differences were found between birds in their ability

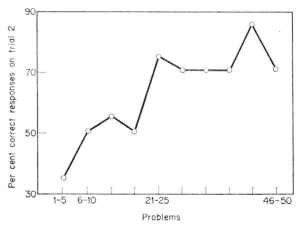

Figure 58. Learning-set formation in young chickens. The graph shows that over the fifty problems there is an increase from 35 per cent to 70 per cent correct responses on the second trial in each problem. (From Plotnik and Tallarico [178].)

to master this problem. Most important, however, was Honigman's analysis of the significance of spacing to the birds. When the distances between grains were altered suddenly from the standard distance used in training, the birds failed but, when the distances were changed gradually, the birds successfully pecked alternate grains. The same effects were found if the distances between grains in a row were made unequal. The birds showed no preference for odd or even grains when pecking the alternate grains in a row of loose grains. One bird after an interruption of five months gave perfect results after ninety-nine trials compared with 550 trials necessary at her introduction to the problem. Nevertheless, using these methods it proved impossible to train birds to peck at every third grain.

Differences in the methods of learning of a single problem by chicks have been shown in chapter 2 from the findings of Munn [167]. In a social animal like the domestic fowl social facilitation might be expected, and a very good example by Smith [212] was described in chapter 4. Chicks in their second week learnt to run down an alley to get food, and the presence of a trained chick improved the running speed of an untrained chick compared to that of a chick running alone or with another untrained chick.

Imprinting

Bateson [21] has defined imprinting as the process which restricts social preferences to a specific class of objects. An imprinted chick not only follows a model but prefers it when given a choice.

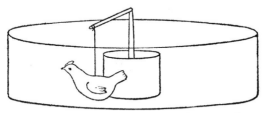

Figure 59. A diagram of a circular track showing the model hen suspended from an arm that can rotate round the track.

Not all chicks can be imprinted to a model in laboratory conditions [96], but such birds might become imprinted to the dam in natural conditions in which the dam would supply a large amount of primary reinforcement by giving the chick warmth and showing it food supplies.

The evidence for a sensitive period is very strong [21], as will be seen below, but in natural and modern agricultural conditions the importance of a sensitive period is likely to be small. For under natural conditions the chick is with the hen from hatching and not suddenly presented with her at a given age. Similarly, a chick reared in a modern poultry unit and imprinted on its companions is in the presence of other chicks from the time it first enters the brooders; before then it will have had auditory and very possibly tactile experience of other chicks in the incubators. Once it is in the brooder its companions become associated with many sorts of reinforcement, such as warmth and food.

Jaynes [122] reported that imprinting, as measured by the retention of the following response ten days after a thirty-minute session with the training model, was greatest in chicks first exposed to the model at thirty-six hours. In rather different conditions, following was at a maximum in chicks aged thirty-six hours which, like Jaynes' chicks had had social experience with other chicks before training [179]. Salzen and Sluckin [193], however, obtained following at five days by chicks which had had social experience but only small experience of the alley and then had seen the model only when it was stationary. Isolation has been found to be effective in prolonging the sensitive period [95, 96]. The end of this period is due to the process of imprinting itself for the chick learns the details of its own environment and as a consequence avoids objects not resembling that environment [20].

The characteristics of a model that attract a chick have been described in the section dealing with the hen-chick relationship. Some ethologists call the stimuli emanating from the hen or model releasers for the following response. Sluckin [209] supports this view that certain newborn or newly-hatched animals have an inherent tendency to approach and follow certain sources of stimulation. Matthews and Hemmings [156] have formulated the initiation and continuation of following in these terms: the moving model elicits attention because of its novelty and this process then induces increased arousal which in the newly-hatched chick is probably very low; the increased arousal leads to locomotion, which is maintained by the negative reinforcement due to reduced arousal on disappearance of the model and also by the possibly positive reinforcement of increased arousal. This hypothesis, if not wholly satisfactory in detail, agrees to some extent with the ideas of Sluckin [209] that imprinting involves perceptual or exposure learning which, he suggests, may include reinforcement through sensory stimulation. Kovach, Fabricius, and Falt [134], too, have emphasized the similarities between perceptual learning and imprinting.

The problem of arousal has been investigated by Pitz and Ross [177]: exposing the chick to an arousing stimulus, such as a loud sound, in the presence of the model enhances imprinting. Arousal outside the imprinting situation was effective at some ages but not at others [223]. Chicks handled in darkness for ten minutes at five hours or nine hours of age were given training sessions for imprinting at twelve and sixteen hours of age and then tested at thirty and fifty-four hours. Handling at five hours, but not at nine hours, was effective in enhancing imprinting. For a full assessment of reinforcement in imprinting and a complete discussion of current ideas the reader is referred to Bateson [21] and Sluckin [209].

7: Stress and Conflict

Stress has been defined as 'that state within a living creature which results from the interaction of the organism with noxious stimuli or circumstances' [246]. Barnett [18] also uses the term to mean a response to any external agency which tends to disturb the homeostatis of the animal. The first definition, embraces the term conflict as used by psychiatrists. Conflict states have been found to be highly correlated with maladjustment or non-adaptive behaviour [165]. Although the recognition of stress or conflict in animals lies in the manifestation of a pathological state or in the performance of apparently non-adaptive behaviour, its recognition is one of the main difficulties in forming objective standards in animal welfare. In this chapter the emphasis will be on how fowls respond behaviourally to various types of stressors.

The main stressors affecting the behaviour of the domestic fowl can be divided into those operating through (a) the physical environment for example, nutritional deficiencies, abnormal temperatures, (b) particular genotype-environmental interactions, (c) the social milieu of the animal. To understand how a stressor works it is convenient to consider whether it acts directly on the physiology of the bird or whether it has its effects primarily through learning processes. Furthermore it is useful to discuss the responses to stressors in terms of whether they are long- or short-term.

Behavioural responses to long-term stressors

Nutritional deficiencies may lead to long-term stress responses, and many believe that feather pecking and cannibalism are the behavioural consequences of nutritional stressors. Although not all feather pecking and cannibalism can be attributed to this type of stressor, some possible examples may be quoted. Schaible, Davidson and Bandemer [195] studied feather pecking and cannibalism by six-to eight-week-old White Leghorn chicks. The chicks were kept on a diet low in protein, crude fibre, and phosphorus. Feather pecking and cannibalism developed in the control flocks kept on this diet but, in the flocks kept on the basal diet with additional casein given at the rate of 4·5 per cent, the percentage of unpecked chicks was significantly higher (figure 60). Siren [207] has given evidence

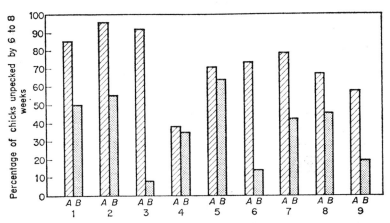

Figure 60. Diet and feather pecking: The percentage of unpecked chicks in 9 groups. A = Low protein diet augmented by casein. B = The same diet without additional casein.
(Adapted from Schaible, Davidson and Bandemer [195].)

of a correlation between arginine deficiency and cannibalism. Fifty-six White Leghorn cockerels were divided into four equal groups at the age of eight weeks. All had been fed on a common starter mash up to that age. All four groups were given a basal mash in which the arginine content was 3·3 per cent of the total protein, two groups received added arginine, and one had extra lysine. The diets and the incidence of cannibalism are shown overleaf.

Miller and Bearse [163] used single-combed White Leghorns aged two to thirty-two weeks, in another study on cannibalism and

Group	Arginine content, % of total protein	No. of birds pecked to death out of 14
1	3·3	9
2	4·5	0
3	3·3 (+ L-lysine)	10
4	6·0	0

pecking. They were divided into sixteen groups. Eight diets were used, each diet fed to two groups. The diets were nearly identical in protein and calcium content. Five of the rations had oats in some form and one had a higher fibre content than the others. Differences in the rate of pecking (the number of birds suffering from canni-balism or feather, toe, comb, or vent pecking) were found between the groups. It was lowest in the groups receiving oat hulls, milled or whole oats, but oat ash or oat hull ash did not have any effect on pecking. The body weights were apparently similar in all the groups but no statistical treatment of body weights is given. In a later experiment these workers [23] again found that White Leghorns fed on oat hulls or oat hull fibre pecked less than White Leghorns fed on the same diet deficient only in oat hulls or oat hull fibre. However, the results with regard to cannibalism were less clear. Neal [169] gives some indication that methionine was effective in eliminating feather pecking in laying Rhode Island Red females, but as he had no proper controls his evidence can only be taken as being suggestive. Creek and Dendy [46], on the contrary, were unable to eliminate cannibalism by White Leghorn cockerels with supplementary DL-methionine. Negative findings however, suggest only that the stock used and other environmental factors can also be important in producing the stress.

The feeding of pellets instead of mash has been suggested as being conducive to feather pecking in that birds consume their food more quickly, become bored, and then feather peck. There appears to be nothing in the literature to support this. In fact, in one example, a severe outbreak of cannibalism occurred in two groups of hens one of which was pellet-fed and the other mash and grain fed [153].

The evidence correlating dietary deficiencies with cannibalism and feather pecking does suggest some interesting questions. How does a dietary deficiency lead to that particular response? A need may be expected to lead to increased activity and responsiveness to stimuli but, if only increased responsiveness to stimuli is involved, how does it lead to pecking of flock mates and not other things in the environment? Does it also lead to increased aggression and hence invariably to social stress? Are learning processes involved? What are the immediate responses to a dietary deficiency? In the prevention of cannibalism and feather pecking, studies on the short-term responses to dietary stressors may be very valuable.

Noise as a stressor in fowls has been only little investigated [215, 216]. It does not appear to influence growth records or the hatchability of eggs in incubators. This, of course, is a crude method of assessing a stressor: other measures are necessary, particularly in view of the litigation that arises over excessive noise.

No gene operates in isolation from the internal or external environment of the animal. Nevertheless certain traits classifiable as non-adaptive and symptomatic of stress, have large genetical components of variance. Among mice, these include whisker eating (D. Faulkner, personal communication) and alcohol preference [161]. There is no unequivocal evidence that feather-pecking is genetically determined for, while Richter [188] stated that it was genetically controlled in his stocks, Dickenson, Kashyap and Lamoureux [50] found only a very low value for its heritability in their stock. Both these papers omit any detailed description of the rearing and maintenance of the birds; hence the findings are open to doubt when the trait is so easily affected by a bird's experience.

Genetic variation among fowls in the utilization of certain vitamins and minerals have been described by Hutt [118]. Furthermore the chick's arginine requirements can be easily changed by selection [119]. Although there was no feather pecking by chicks from the line with the high requirement for arginine, traits of this sort could in certain environments lead to non-adaptive behaviour.

The chicken is a highly social animal, and interruption of social behaviour may act as a stressor. In several experiments, some of which have been discussed earlier, chicks have been isolated from conspecifics from hatching until adulthood. Such birds show disturbed behaviour during isolation; for example, chasing their tails and exhibiting excessive aggressiveness. When released into the social environment the birds usually adapt themselves [29, 252], although there may be 'withdrawal' or atypical aggressive behaviour at the beginning. Inhibition of sexual behaviour in Red Junglefowl males isolated until ten or more months of age has been reported [136]. Whether this inhibition was due to early

(a)   (b)

Figure 61. A Burmese Red Junglefowl cockerel that has been reared in isolation shows aggressive behaviour to a crouching female. (From Kruijt [136].)

isolation is difficult to assess, for inhibition may also be manifested through conditioning in adulthood.

Guhl, Collias, and Allee [85] found that when several males were placed together in a pen of hens, the dominant cock did most of the mating and prevented the others from mating. When the dominant male was removed the other cocks did not tread often, although they had been sexually vigorous before the peck-order developed. The authors concluded that the inferior cocks had been so conditioned as to be 'psychologically castrated'.

Many 'perversions' in the sexual behaviour of the fowl have been described by several authors [68, 98, 201]. All these birds were prevented from making normal social bonds during the sensitive period for imprinting, either by being kept in isolation, or by being imprinted to other species so that they directed their adult sexual behaviour to the latter.

Social stress among rodents can lead to adrenal hypertrophy [18]. Similarly in a comparison between flocks of White Leghorns housed at 1200 cm^2 per bird and others at 3600 cm^2 per bird, but with feeding, watering, and nest space otherwise equal, Siegel [204] found that during 196 days egg production was reduced at the higher density and adrenal weights increased. In a later study [205] no correlation was found between the fighting success of thirteen-month-old cocks and their adrenal weights, nor was any correlation found between adrenal weight and peck order position six days after the peck-order was formed. In younger cocks no correlation between peck-order position and adrenal gland weight was found after an experiment lasting twenty-one days. However, when Leghorn males aged three-and-a-half months were placed for four hours daily for a period of twenty-two days in pens with strange males, their left adrenals were heavier than those of control males that had been kept in stable groups. Their plasma cholesterol levels were also lower than those of the controls.

Clearly more information is needed on social stress involving longer periods than Siegel and Siegel have employed. Perhaps we are wrong to expect a straightforward correlation between the state of the adrenal glands of a fowl and its peck-order position: for it may be just as 'stressful' to maintain a high position as to be kept in a low position. It is necessary for us also to know more about the responses of birds to new surroundings and changes in husbandry practices, and the retention of conditioned avoidance responses together with the degree of generalization of these responses. Similarly, detailed studies on the development of apparently non-adaptive behaviour to long-term stressors, such as dietary deficiencies, are needed.

However, it must not be assumed that in the long term all stress is entirely bad, for a certain amount is probably even beneficial. An example is provided by a very interesting experiment by Gross and Colman [83]. Fowls were kept in cages in groups of six, and some birds were regularly moved to strange peck-orders for short periods while others remained as residents but received visitors to their peck-order. A third class were not moved, received no visitors, but were regularly handled. Resistance to *Escherichia coli* was greatest in the visitors and least in the third class of birds.

Behavioural responses to short-term stressors

Among the short-term responses to stressors are 'displacement' activities, the seemingly irrelevant activities that suddenly appear in the animals' responses during conflict or thwarting. As we saw earlier, displacement activities are very common in social situations and in fact most of the agonistic and courtship displays show elements of ambivalence, although some may now be emancipated from their original motivating mechanisms.

Wood-Gush [250] found that most of the normal courtship displays were performed by cockerels physically thwarted from reaching the hens. However, the observation periods may have been too short for the males to feel really thwarted. Guiton and Wood-Gush [100] studied the behaviour of hungry hens when thwarted in their attempts to get food. The birds which were of Rhode Island Red extraction, were trained to expect food in the test cage. In the first experiment the birds, which were tested individually, were thwarted by placing a pane of clear glass over the food dish. Before the thwarting test they had been deprived of food for twenty-four hours. Two control situations were employed. The birds were individually placed in the test cage for the same period at the same time of day. In one situation the birds were not hungry and there was no food available; in the other they were hungry and had access to food. The immediate response to thwarting was escape behaviour, which was not present in the control situation, preening, which is common as a displacement activity, was not very frequent. However, by the fourth test the number of escapes had declined, while preening increased in frequency between the first and last tests. More detailed analysis of the findings showed that seventy-six per cent of the behaviour patterns containing feeding elements in the first tests were closely associated with escape, whereas in the last test such behaviour patterns were more closely related to preening and other related movements, commonly considered to be displacement activities. Another common element of behaviour under thwarting was redirected pecking, which initially was closely related in time to escape behaviour and in later tests became more associated with grooming.

Escape is probably a primary reaction to thwarting which becomes replaced by other activities such as preening. The repeti-

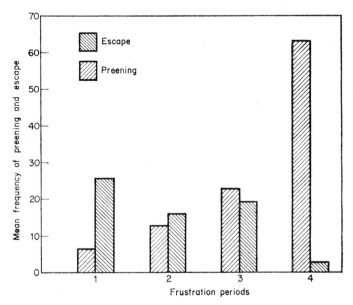

Figure 62. Histogram showing the mean relative frequencies of escape behaviour and preening by three hungry hens when thwarted on four occasions. Note the decline in escape behaviour and increase in preening from the first thwarting experience to the fourth. (Adapted from Guiton and Wood-Gush [100].)

tion of thwarting may lead to a reduction of its aversive effects as well as a reduction in stimulation to feed. As attempts to feed were extinguished and the bird's attention was diverted, irrelevant activities occurred. Escape could be interpreted as trying to leave the cage to get to the food and so might not be classified as irrelevant to the situation.

In a second type of test two of the same birds were thwarted from feeding by delivering a blast of oxygen under high pressure every time they fed. This method differed from the first in that it introduced an obvious element of fear into the situation and thus induced a conflict between two 'drives'. Only one control situation was used, in which the bird was not hungry and no food was present. Again the experimental and control situations were performed at the same time of day. Under this type of thwarting the most common behaviour pattern was to-and-fro movement. This was far more frequent than escapes or preening, which remained at very low frequency in all the experimental situations. In this situation it might have been expected from previous ethological theories that displacement activities would have been frequent in periods when the approach and avoidance tendencies were in equilibrium. However, even when the frequency of to-and-fro movements declined, the conflict seemed to remain very intense, for the bird stood moving its head towards, and away from, the container. It apparently remained fixated on the food stimulus although too frightened to approach. Apparently attention to other stimuli remained low and no irrelevant activities occurred. In this respect these birds followed the pattern of the Barbary doves studied by McFarland [162] under thwarting and conflict conditions. He distinguished two types of stationary posture in these situations: a Stationary Attentive Posture (SAP) in which the bird is relaxed and alert, and a Stationary Ambivalent Posture (SAV) in which it fixates on the stimulus object and seems to be in a state of balance between the two tendencies, to approach and to retreat. He reported the latter to be more common in conflict situations, whereas SAP was more frequent during thwarting; a high SAP : SAV ratio was accompanied more often by displacement activities than was a low ratio.

The daily life of the average fowl is very likely to bring it thwarting, for competition around the food trough may be intense even in settled peck-orders for birds of fairly high status [255]. It would be of interest therefore to know a great deal more about how the bird deals with thwarting imposed by social factors. Furthermore, one would like to know whether this type of thwarting, which is likely to be very common, has any long term effects, or whether the short-term responses serve to alleviate the position. An inventory of the fowl's behavioural responses in conflict situations and to thwarting of each of its main drives, together with telemetry of certain physiological states during these periods, would tell us whether certain overt behavioural responses are correlated with internal signs of stress. Finally, long term studies of this sort would enable us to assess stress more objectively and arrive at sounder standards of animal welfare.

Bibliography

1 ALLEE, W. C., COLLIAS, N. E. and LUTHERMAN, C., 1939. 'Modification of the social order in flocks of hens by the injection of testosterone propionate': *Physiol. Zool.* 12, 412–40.

2 ALLEE, W. C. and COLLIAS, N. E., 1940. 'The influence of oestradiol on the social organization of flocks of hens': *Endocrinology* 27, 87–94.

3 ALLEE, W. C., COLLIAS, N. E. and BEEMAN, E., 1940. 'The effect of thyroxin on the social order in flocks of hens': *Endocrinology* 27, 827–35.

4 ALLEE, W. C., FOREMAN, D., BANKS, E. M., and HOLABIRD, C. H., 1955. 'Effects of androgen on dominance and subordinance in six common breeds of *Gallus Gallus*': *Physiol. Zoöl.* 28, 89–115.

5 ALPERT, M., SCHEIN, M. W., BECK, C. H. and WARREN, J. M., 1962. 'Learning-set formation in young chicks': *Am. Zool.* 2, 263.

6 ANDREW, R. J., 1956. 'Normal and irrelevant toilet behaviour in *Emberiza spp.*': *Brit. J. Anim. Behav.* 4, 85–91.

7 ANDREW, R. J., 1966. 'Precocious adult behaviour in the young chick': *Anim. Behav.* 14, 485–500.

8 ANDREW, R. J., 1967. 'Intracranial self-stimulation in the chick': *Nature* Lond. 213, 847–8.

9 ANDREW, R. J., 1964. 'Vocalization in chicks, and the concept of "stimulus contrast" ': *Anim. Behav.* 12, 64–76.

10 ARMINGTON, J. C. and THIEDE, F. C., 1956. 'Electroretinal demonstration of a Purkinje shift in the chicken eye': *Am. J. Physiol.* 186, 258–62.

11 ASCHOFF, J. and VON HOLST, D., 1958. 'Schlafplatzflüge der Dohle, *Corvus monedula* L': *Proc. XIIth Internat. Ornithol. Congr. Helsinki* 55–70 (1960).

12 ASCHOFF, J., 1963. 'Comparative physiology: diurnal rhythms': *Ann. Rev. Physiol.* 25, 518–600.

13 ASCHOFF, J., 1964. 'Survival value of diurnal rhythms': *Symp. Zool. Soc. Lond.* 13, 79–98.

14 BACON, H. R., WARREN, J. M. and SCHEIN, M. W., 1962. 'Nonspatial reversal learning in chickens'. *Anim. Behav.* 10, 239–43.

15 BAEUMER, E., 1955. 'Lebensart des Haushuhns': *Z. Tierpsychol.* 12, 387–401.

16 BAEUMER, E., 1962. 'Lebensart des Haushuhns, dritter Teil – über seine Laute und allgemeine Ergänzungen.': *Z. Tierpsychol.* 19, 394–416.

17 BANKS, E. M. and ALLEE, W. C., 1957. 'Some relations between flock size and agonistic behavior in domestic hens': *Physiol. Zoöl.* 30, 255–68.

18 BARNETT, S. A., 1964. 'Social stress', in 'Viewpoints in biology'. Ed. Carthy, J. D. and Duddington, C. L. London, Butterworths.

19 BASTOCK, M., 1967. 'Courtship: a zoological study': Heinemann Educational Books, London.

20 BATESON, P. P. G., 1964. 'Effect of similarity between rearing and testing conditions on chicks' following and avoidance responses': *J. comp. physiol. Psychol.* 57, 100–3.

21 BATESON, P. P. G., 1966. 'The characteristics and context of imprinting': *Biol. Rev.* 41, 177–220.

22 BAYER, E., 1929. 'Beitrage zur Zweikomponentheorie des Hungers (Versuche mit Hühnern)': *Z. Psychol.* 112, 1–54.

23 BEARSE, G. E., MILLER, V. L. and McCLARY, C. F., 1940. 'The cannibalism preventing properties of the fiber fraction of oats': *Poult. Sci* 19, 210–15.

24 BEEBE, W., 1918. A monograph of the Pheasants: Witherby, New York.

25 BENNER, J., 1938. 'Untersuchungen uber Raumwahrnehmung': *Z. wiss. Zool.* 151, 382–444.

26 BINGHAM, H. C., 1914. 'A definition of form': *J. anim. Behav.* 4, 136–41.

27 BREMOND, J. C., 1963. 'Acoustic behavior of birds', in *Acoustic Behavior of Animals*: Busnel, R. G. (Ed.). Elsevier, London.

28 BROWN, J. L. and HUNSPERGER, R. W., 1963. 'Neuro-ethology and the motivation of agonistic behaviour': *Anim. Behav.* 11, 439–48.

29 BRUCKNER, G. H., 1933. 'Untersuchungen zur Tiersoziologie, insbesondere zur Auflösing der Familie': *Z. Psychol.* 128, 1–105.

30 BURROWS, W. H. and BYERLY, T. C., 1938. 'The effect of certain groups of environmental factors upon the expression of broodiness': *Poult. Sci.* 17, 324–30.

31 CAMPBELL, B. A., SMITH, N. F., MISANIN, J. R. and JAYNES, J., 1966. 'Species differences in activity during hunger and thirst': *J. comp. physiol. Psychol.* 61, 123–7.

32 CAPRETTA, P. J., 1961. 'An experimental modification of food preference in chickens': *J. comp. physiol. Psychol.* 54, 238–42.

33 CARBAUGH, B. T., SCHEIN, M. W. and HALE, E. B., 1962. 'Effects of morphological variations of chicken models on sexual responses of cocks': *Anim. Behav.* 10, 235–8.

34 CHANCE, M. R. A., 1962. 'An interpretation of some agonistic postures; The role of 'cut off' acts and postures': *Symp. Zool. Soc. Lond.* 8, 71–89.

35 CHATTOCK. A. P: and GRINDLEY, G. C., 1931. 'The effect of change of reward on learning in chickens': *Brit. J. Psychol.* 22, 62–6.

36 CHATTOCK, A. P. and GRINDLEY, G. C., 1933. 'The effect of delayed reward on the maze performance of chickens': *Brit. J. Psychol.* 23, 382–8.

37 COLE, L. W., 1911. 'The relation of strength of stimulus to rate of learning in the chick': *J. anim. Behav.* 1, 111–24.

38 COLLIAS, N. E., 1943. 'Statistical analysis of factors which make for success in initial encounters between hens': *Am. Nat.* 77, 519–38.

39 COLLIAS, N. E., 1944. 'Aggressive behavior among vertebrate animals': *Physiol. Zoöl.* 17, 83–123.

40 COLLIAS, N. E., 1950. 'Hormones and behavior with special reference to birds and the mechanisms of hormone action', in *A symposium on steroid hormones*. Ed. Gordon, E. University of Wisconsin Press, 277–329.

41 COLLIAS, N. E., 1952. 'The development of social behavior in birds': *Auk* 69, 127–59.

42 COLLIAS, N. E. and COLLIAS, E. C., 1967. 'A field study of the Red Jungle Fowl in North-central India': *Condor* 69, 360–86.

43 COLLIAS, N. E. and JOOS, M., 1953. 'The spectrographic analysis of sound signals of the domestic fowl': *Behaviour* 5, 175–88.

44 COLTHERD, J. B., 1966. 'The domestic fowl in Ancient Egypt': *Ibis* 108, 217–23.

45 CRAIG, J. V., ORTMAN, L. L. and GUHL, A. M., 1965. 'Genetic selection for social dominance ability in chicks': *Anim. Behav.* 13, 114–31.

46 CREEK, R. D. and DENDY, M. Y., 1957. 'The relationship of cannibalism and methionine': *Poult. Sci.* 36, 1093–4.

47 CRUZE, W. W., 1935. 'Maturation and learning in chicks': *J. comp. Psychol.* 19, 371–410.

48 DAVIS, D. E. and DOMM, L. V., 1943. 'The influence of hormones on the sexual behavior of the fowl'. *Essays in Biology*, 171–81. University of California Press.

49 DAWKINS, R., 1968. 'The ontogeny of a pecking preference in domestic chicks': *Z. Tierpsychol.* 25, 170–86.

50 DICKENSON, G., KASHYAP, T. and LAMOREUX, W. F., 1961. 'Heritable variation in pecking behaviour of chickens': *Poult. Sci.* 40, 1394–5.

51 DOMM, L. V., 1924. 'Sex reversal following ovariotomy in the fowl': *Proc. Soc. exp. Biol. N.Y.* 22, 28–35.

52 DOMM, L. V., 1929. 'The effects of bilateral ovariotomy in the Brown Leghorn fowl': *Biol. Bull.* 56, 459–97.

53 DOMM, L. V., 1930. 'Artificial insemination with motile sperm from ovariotomized fowl': *Anat. Rec.* 47, 297.

54 DOMM, L. V., 1931. 'A demonstration of equivalent potencies of right and left testis-like gonads in the ovariectomized fowl': *Anat. Rec.* 49, 211–49.

55 DOMM, L. V. and BLIVASS, B. B., 1947. 'Induction of male copulatory behaviour in the Brown Leghorn hen': *Proc. Soc. exp. Biol. N.Y.* 66, 418–19.

56 DOMM, L. V. and DAVIS, D. E., 1948. 'The sexual behaviour of the intersexual domestic fowl': *Physiol. Zoöl.* 21, 14–31.

57 DOUGLIS, M. B., 1948. 'Social factors influencing the hierarchies of small flocks of the domestic hen; interactions between resident and part-time members of organized flocks': *Physiol. Zoöl.* 21, 147–82.

58 DRYE, K. J., GILBREATH, J. C. and MORRISON, R. D., 1959. 'The effects of reserpine on chicken males on range': *Poult. Sci.* 38, 781–6.

59 EISNER, E., 1960. 'The relationship of hormones to the reproductive behaviour of birds, referring especially to parental behaviour: a review': *Anim. Behav.* 8, 155–79.

60 ENGELMAN, C., 1940. 'Versuche über die Beliebheit einiger Getreide-Arten beim Hühn': *Z. vergl. Physiol.* 27, 525–44.

61 ENGELMANN, C., 1951. 'Beiträge zum Gedachtnis des Hühns': *Z. Tierpsychol.* 8, 110–21.

62 FANTZ, R. L., 1957. 'Form preferences in newly hatched chicks': *J. comp. physiol. Psychol.* 50, 422–30.

63 FANTZ, R. L., 1959. 'Response to horizontality by Bantam chickens in level and tilted room': *Psychol. Rec.* 9, 61–6.

64 FELDMAN, S. E., LARSSON, S., DIMICK, M. K. and LEPKOVSKY, S., 1957. 'Aphagia in chickens': *Am. J. Physiol.* 191, 259–61.

65 FISCHEL, W., 1927. 'Beiträge zum soziologie des Haushuhns': *Biol. Zbl.* 47, 678–95.

66 FISCHER, G. J., 1966. 'Auditory stimuli in imprinting': *J. comp. physiol. Psychol.* 61, 271–3.

67 FISHER, H. and WEISS, H. S., 1956. 'Feed consumption in relation to dietary bulk and energy level: the effect of surgical removal of the crop': *Poult. Sci.* 35, 418–23.

68 FISHER, A. E. and HALE, E. B., 1957. 'Stimulus determinants of sexual and aggressive behavior in male domestic fowl': *Behaviour* 10, 309–23.

69 FISHMAN, R. and TALLARICO, R. B., 1961a. 'Studies of visual depth perception: I. Blinking as an indicator response in prematurely hatched chicks': *Percept. Mot. Skills* 12, 247–50.

70 FISHMAN, R. and TALLARICO, R. B., 1961b. 'Studies of visual depth perception: II. Avoidance reaction as an indicator response in chicks': *Percept. Mot. Skills* 12, 251–7.

71 FOREMAN, D. and ALLEE, W. C., 1959. 'A correlation between posture stance and outcome in paired contests of domestic hens': *Anim. Behav.* 7, 180–8.

72 FOSTER, E. S. and HEFFER, E. H., 1941. 'Columella: De Re Rustica': Heinemann, London.

73 FUERST, W. F. and KARE, M. R., 1962. 'The influence of pH on fluid tolerance and preferences': *Poult. Sci.* 41, 71–7.

74 FUNK, E. M., 1932. 'Can the chick balance its ration?': *Poult. Sci.* 11, 94–7.

75 GILBERT, A. B. and WOOD-GUSH, D. G. M., 1965. 'The control of the nesting behaviour of the domestic hen. III. The effect of cocaine in the post-ovulatory follicle': *Anim. Behav.* 13, 284–5.

76 GILMAN, I. T., MARCUSE, F. L. and MOORE, A. U., 1950, 'Animal hypnosis: a study in the induction of tonic immobility in chickens': *J. comp. physiol. Psychol.* 43, 99–111.

77 GOODALE, H. D., 1916. 'Notes on the behaviour of capons when brooding chicks': *J. anim. Behav.* 6, 319–24.

78 GRAHAM, J. C., 1934. 'Individuality of pullets in balancing rations': *Poult. Sci.* 13, 34–9.

79 GRAHAM, W. R., 1932. 'Can we learn anything from a free choice of feeds as expressed by chickens?': *Poult. Sci.* 11, 365–6.

80 GRAY, P. H., 1957. 'Irrelevant cue learning in the chick': *Psychol. Rep.* 3, 345–52.

81 GRINDLEY, G. C., 1927. 'Experiments on the "direction of association" in young chickens': *Brit. J. Psychol.* 17, 210–21.

82 GRINDLEY, G. C., 1929. 'Experiments on the influence of the amount of reward on learning in young chicks': *Brit. J. Psychol.* 20, 173–80.

83 GROSS, W. B. and COLMANO, G., 1967. 'Further studies on the effects of social stress in the resistance to the infection with Escherichia coli': *Poult. Sci.* 46, 41–6.

84 GUHL, A. M. and ALLEE, W. C., 1944. 'Some measureable effects of social organization in flocks of hens': *Physiol. Zoöl.* 17, 320–47.

85 GUHL, A. M., COLLIAS, N. E. and ALLEE, W. C., 1945. 'Mating behavior in the social hierarchy in small flocks of White Leghorns': *Physiol. Zoöl.* 18, 365–90.

86 GUHL, A. M., 1948. 'Unisexual mating in a flock of White Leghorn hens': *Trans. Kans. Acad. Sci.* 51, 107–11.

87 GUHL, A. M., 1950a. 'Heterosexual dominance and mating behaviour in chickens': *Behaviour* 2, 106–19.

88 GUHL, A. M., 1950b. 'Social dominance and receptivity in the domestic fowl': *Physiol. Zoöl.* 23, 361–6.

89 GUHL, A. M., 1953. 'The social behavior of the domestic fowl': *Tech. Bull. Agric. exp. Sta.* 73 Kans. State College, 48.

90 GUHL, A. M. and ORTMAN, L. L., 1953. 'Visual patterns in the recognition of individuals among chickens': *Condor* 55, 287–98.

91 GUHL, A. M., 1958. 'The development of social organization in the domestic chick': *Anim. Behav.* 6, 92–111.

92 GUHL, A. M., CRAIG, J. V. and MUELLER, C. D., 1960. 'Selective breeding for aggressiveness in chickens': *Poult. Sci.* 39, 970–80.

Bibliography

93 GUHL, A. M., 1962. 'The Behavior of Chickens', in *The Behavior of Domestic Animals*. Ed. Hafez, E. S. E., Bailliere, Tindall and Cox, London.

94 GUHL, A. M., 1964. 'Psychophysiological interrelations in the social behavior of chickens': *Psychol. Bull.* 61, 277–85.

95 GUITON, P., 1958. 'The effect of isolation on the following response of Brown Leghorn Chicks': *Proc. R. phys. Soc. Edinb.* 27, 9–14.

96 GUITON, P., 1959. 'Socialization and imprinting in Brown Leghorn Chicks': *Anim. Behav.* 7, 26–34.

97 GUITON, P., 1961. 'The influence of imprinting on the agonistic and courtship responses of the Brown Leghorn cock': *Anim. Behav.* 9, 167–77.

98 GUITON, P., 1962. 'The development of sexual responses in the domestic fowl in relation to the concept of imprinting': *Symp. Zool. Soc. Lond.* No. 8, 227–34.

99 GUITON, P., 1966. 'Early experience and sexual object-choice in the Brown Leghorn': *Anim. Behav.* 14, 534–8.

100 GUITON, P. and WOOD-GUSH, D. G. M., 1967. 'Studies on thwarting in the domestic fowl': *Rev. comp. anim.* 5, 1–23.

101 GUNTHER, W. C. and JONES, R. K., 1961a. 'Effect of non-optimally high incubation temperatures on T-maze learning in the chick': *Proc. Indiana Acad. Sci.* 71, 327–33.

102 GUNTHER, W. C. and JONES, R. K., 1961b.'Effect of environmental stress on chick weight': *Proc. Indiana Acad. Sci.* 71, 385–98.

103 GUYOMARC'H J-Ch., 1962. 'Contribution a l'étude du comportement vocal du poussin de "Gallus domesticus" ': *J. Psychol. norm. path.* 3, 283–306.

104 HALE, E. B., 1948. 'Observations on the social behaviour of hens following debeaking': *Poult. Sci.* 27, 591–2.

105 HALE, E. B., 1957. 'Breed recognition in the social interactions of domestic fowl': *Behaviour* 10, 240–54.

106 HALPERN, B. P., 1962. 'Gustatory nerve responses in the chicken': *Am. J. Physiol.* 203, 541–4.

107 HALPERN, B. P., 1967. 'Some relationships between electro-physiology and behavior in taste', in *The chemical senses and nutrition*. Eds. Kare, M. R. and Maller. O. Johns Hopkins University Press, Baltimore.

108 HAMILTON, J. B., 1938. 'Precocious masculine behavior following administration of synthetic male hormone substances': *Endocrinology* 23, 53–7.

109 HELLWALD, H., 1931. 'Untersuchungen über Triebstarken bei Tieren': *Z. Psychol.* 123, 94–141.

110 HESS, E. H., 1950. 'Development of the chick's responses to light and shade cues of depth': *J. comp. physiol. Psychol.* 43, 112–22.

111 HESS, E. H., 1956a. 'Space perceptions in the chick': *Sci. Am.* 195, 71–8.

112 HESS, E. H., 1956b. 'Natural preference of chicks and ducklings for objects of different colours': *Psychol. Rep.* 2, 477–83.

113 HONIGMAN, H., 1942. 'The alternation problem in animal psychology: experiments with fowl': *J. exp. Biol.* 19, 141–57.

114 HUTCHINSON, J. C. D. and TAYLOR, W. W., 1962a. 'Motor co-ordination of pecking fowls': *Anim. Behav.* 10, 55–61.

115 HUTCHINSON, J. C. D. and TAYLOR, W. W., 1962b. 'Effect of exposure to heat on the motor co-ordination of pecking fowls': *Anim. Behav.* 10, 62–6.

116 HUTCHINSON, J. C. D. and TAYLOR, W. W., 1962c. 'Mechanics of pecking grain': *World's Poult. Congr.* XIIth, 112–16.

117 HUTT, F. B., 1949. 'Genetics of the Fowl': McGraw-Hill, New York, Toronto, and London.

118 HUTT, F. B., 1961. 'Genetic variation in the utilisation of riboflavin, thiamine and other nutrients': *Ann. N.Y. Acad. Sci.* 91, 659–66.

119 HUTT, F. B. and NESHEIM, M. C., 1966. 'Changing the chick's requirement of arginine by selection': *Can. J. Genet. Cytol.* 8, 251–9.

120 JAMES, J. W. and FOENANDER, F., 1961. 'Social behaviour studies on domestic animals. I. Hens in laying cages': *Aust. J. agric. Res.* 12, 1239–52.

121 JAMES, H., 1959. 'Flicker: an unconditional stimulus for imprinting': *Can. J. Psychol.* 13, 59–67.

122 JAYNES, J., 1957. 'Imprinting: The interaction of learned and innate behavior. II. The critical period': *J. comp. physiol. Psychol.* 50, 6–10.

123 JUKES, L. L., 1938. 'The selection of diet in chicks': *J. comp. Psychol.* 26, 135–56.

124 KARE, M. R., BLACK, R. and ALLISON, E. G., 1957. 'The sense of taste in the fowl': *Poult. Sci.* 36, 129–38.

125 KARE, M. R. and MEDWAY, W., 1959. 'Discrimination between carbohydrates by the fowl': *Poult. Sci.* 38, 1119–26.

126 KARE, M. R. and SCOTT, M. L., 1962. 'Nutritional value and feed acceptability': *Poult. Sci.* 41, 276–8.

127 KATZ, D. and REVESZ, G., 1909. 'Experimentall-psychologische Unter-suchungen mit Hühnern': *Z. Psychol.* 50, 93–116.

128 KATZ, D. and KELLER, H. H., 1924. 'Das Zielen bei Tieren (Versuche mit Hühnern)': *Z. Psychol.* 95, 27–35.

129 KING, M. G., 1965a. 'Disruptions in the pecking order of cockerels concomitant with degrees of accessibility to feed': *Anim. Behav.* 13, 504–6.

130 KING, M. G., 1965b. 'The effect of social context on dominance capacity of domestic hens': *Anim. Behav.* 13, 132–3.

131 KITCHELL, R. L., STRÖM, L. and ZOTTERMAN, Y., 1959. 'Electro-physiological studies of thermal and taste reception in chickens and pigeons': *Acta. physiol. scand.* 46, 133–51.

132 KOMAI, T., CRAIG, J. V. and WEARDEN, S., 1959. 'Heritability and repeatability and social aggressiveness in the domestic chicken': *Poult. Sci.* 38, 356–9.

133 KONISHI, M., 1963. 'The role of auditory feed-back in the vocal behaviour of the domestic fowl': *Z. Tierpsychol.* 20, 349–67.

134 KOVACH, J. K., FABRICIUS, E. and FÄLT, L., 1966. 'Relationship between imprinting and perceptual learning': *J. comp. physiol. Psychol.* 61, 449–54.

135 KRUIJT, J. P., 1962. 'Imprinting in relation to drive interactions in Burmese Red Jungle Fowl': *Symp. Zool. Soc. Lond.* 8, 219–26.

136 KRUIJT, J. P., 1964. 'Ontogeny of social behaviour in Burmese Red Jungle Fowl (*Gallus gallus spadiceus*)': Brill, Leiden.

137 KUIPER, J. W. and UBBELS, P., 1951. 'A biological study of natural incubation and its application to artificial incubation': *Proc. Wld's. Poult. Congr.* IXth Section 1, 105-12.

138 KUO, Z. Y., 1932. 'Ontogeny of embryonic behavior in Aves. I. The chronology and general nature of the behavior of the chick embryo': *J. exp. Zool.* 61, 395–430.

139 LAKE, P. E. and WOOD-GUSH, D. G. M., 1956. 'Diurnal rhythms in semen yields and mating behaviour in the domestic cock': *Nature* Lond. 178, 853.

140 LAKE, P. E., 1957. 'Fowl semen as collected by the massage method': *J. agric. Sci.* Camb. 49, 120–6.

141 LAKE, P. E. and EL JACK, M. H., 1964. 'The origin and composition of fowl semen': in *Physiology of the domestic fowl*. Symp. No. 1 British Egg Marketing Board. Ed. Horton-Smith, C. and Amoroso, E. C., Oliver & Boyd, Edinburgh and London.

142 LANE, H., 1960. 'Control of vocal responding in chickens': *Science* 132, 37–8.

143 LASHLEY, K. S., 1916. 'The colour vision of birds. I. The spectrum of the domestic fowl': *J. anim. Behav.* 6, 1–26.

144 LEPKOVSKY, S. and YASUDA, M., 1966. 'Hypothalamic lesions, growth and body composition of male chickens': *Poult. Sci.* 45, 582–8.

145 LEPKOVSKY, S. and YASUDA, M. 1967. 'Adipsia in chickens': *Physiol. Behav.*, 2, 45–7.

146 LILL, A. and WOOD-GUSH, D. G. M., 1965. 'Potential ethological isolating mechanisms and assortative mating in the domestic fowl': *Behaviour* 25, 16–44.

147 LILL, A., 1966. 'Some observations on social organization and non-random mating in captive Burmese Red Jungle Fowl (*Gallus gallus spadiceus*)': *Behaviour* 26, 228–42.

148 LINDENMAIER, P. and KARE, M., 1959. 'The taste end-organs of the chicken': *Poult. Sci.* 38, 545–50.

149 LLOYD MORGAN, C., 1896. 'The habit of drinking in young chicks': *Science* 3, 900.

150 LONG, E. and GODFREY, G. F., 1952. 'The effect of dubbing, environmental temperature and social dominance on mating activity and fertility in domestic fowl': *Poult. Sci.* 31, 665–73.

151 LORENZ, F. W., 1953. 'The use of estrogens for fattening poultry': *Nat. Acad. Sci.–Nat. Res. Council* Publ. 266, 5–17.

152 LOWE, P. R., 1934. 'A further note bearing on the data when the domestic fowl was first known to the ancient Egyptians': *Ibis* 4, 378–82.

153 MACINTYRE, T. M. and JENKINS, M. H., 1955. 'Effect of different feeding methods on the efficiency of egg production': *Poult. Sci.* 34, 376–83.

154 MAIER, R. A., 1964. 'The role of the dominance-submission ritual in social recognition of hens': *Anim. Behav.* 12, 59.

155 MARLER, P., KREITH, M. and WILLIS, E., 1962. 'An analysis of testosterone-induced crowing in young domestic cockerels': *Anim. Behav.* 10, 48–54.

156 MATTHEWS, W. A. and HEMMINGS, G., 1963. 'A theory concerning imprinting': *Nature* Lond. 198, 1183–4.

157 MCBRIDE, G., 1958. 'Relationship between aggressiveness and egg production in the domestic hen': *Nature* Lond. 181, 858.

158 MCBRIDE, G. and FOENANDER, F., 1962. 'Territorial behaviour in flocks of domestic fowls': *Nature* Lond. 194, 102.

159 MCBRIDE, G., JAMES, J. W. and SHOFFNER, R. N., 1963. 'Social forces determining spacing and head orientation in a flock of domestic hens': *Nature* Lond. 197, 1272–3.

160 MCBRIDE, G., PARER, I. P. and FOENANDER, F., 1966. 'The social organizations and behaviour of the feral domestic fowl': unpublished communication.

161 MCCLEARN, G. E. and RODGERS, D. A., 1961. 'Genetic factors in alcohol preference of laboratory mice': *J. comp. physiol. Psychol.* 54, 116–19.

162 MCFARLAND, D. J., 1966. 'On the causal and functional significance of displacement activities': *Z. Tierpsychol.* 23, 217–35.

163 MILLER, M. W. and BEARSE, G. E., 1938. 'The cannibalism preventing properties of oat hulls': *Poult. Sci.* 17, 466–71.

164 MILLER, N. E., 1957. 'Experiments on motivation. Studies combining psychological, physiological and pharmacological techniques': *Science* 126, 1271–8.

165 MILLER, N. E., 1961. 'Some recent studies of conflict behavior and drugs': *Am. Psychol.* 16, 12–24.

166 MOORE, C. A. and ELLIOTT, R., 1946. 'Numerical and regional distribution of the taste buds on the tongue of the bird': *J. comp. Neurol.* 84, 119–31.

167 MUNN, N. L., 1931. 'The relative efficiency of form and background in the chicks discrimination of visual patterns': *J. comp. Psychol.* 12, 41–75.

168 NASH, T. L., WARREN, J. M. and SCHEIN, M. W., 1961. 'Effects of time and interpolated problems upon retention in chickens': *Am. Zool.* 1, 377–8.

169 NEAL, W. M., 1956. 'Cannibalism, pick-outs and methionine': *Poult. Sci.* 35, 10–13.

170 OOKAWA, T. and TAKENAKA, S., 1967. 'Depth-encephalograms of the adult chicken's forebrain during behavioural wakefulness and sleep': *Poult. Sci.* 46, 769–71.

171 PADILLA, S. G., 1935. 'Further studies on the delayed pecking of chicks': *J. comp. Psychol.* 20, 413–43.

172 PALMGREN, P., 1949. 'On the diurnal rhythm of activity and rest in birds': *Ibis* 91, 561–76.

173 PARKER, J. E., MCKENZIE, F. F. and KEMPSTER, H. L., 1940. 'Observations on the sexual behaviour of New Hampshire males': *Poult. Sci.* 19, 191–7.

174 PARKER, J. E. and BERNIER, P. E., 1950. 'Relation of male to female ratio in New Hampshire breeder flocks to fertility of eggs': *Poult. Sci.* 29, 377–80.

175 PENQUITE, R., CRAFT, W. A. and THOMSON, R. B., 1929. 'Variation in activity and production of spermatozoa by White Leghorn males': *Poult. Sci.*, 9, 247–56.

176 PETERS, J. P., 1913. 'The Cock': *J. Am. orient. Soc.* 33, 363–401.

177 PITZ, G. F. and ROSS, R. B., 1961. 'Imprinting as a function of arousal': *J. comp. physiol. Psychol.* 54, 602–4.

178 PLOTNIK, R. J. and TALLARICO, R. B., 1966. 'Object-quality learning-set formation in the young chicken': *Psychon. Sci.* 5, 195–6.

179 POLT, J. M. and HESS, E. H., 1966. 'Effects of social experience on the following response in chicks': *J. comp. physiol. Psychol.* 61, 268–70.

180 POTTER, J. H., 1949. 'Dominance relations between different breeds of domestic hens': *Physiol. Zoöl.* 22, 261–80.

181 RACKHAM, H., 1940. 'Pliny: Natural History Books 8–11': Heinemann, London.

182 RAMSAY, A. O., 1953. 'Variations in the development of broodiness in fowl': *Behaviour* 5, 51–7.

183 REICHNER, H., 1925. 'Über farbige Umstimmung (Sukzessivkontrast) und Momentadaptation der Hühner': *Z. Psychol.* 96, 68–75.

184 REVESZ, G., 1922. 'Tierpsychologische Untersuchungen': *Z. Psychol.* 88, 130–2.

185 REVESZ, G., 1924. 'Experiments on animal perception': *Brit. J. Psychol.* 14, 387–414.

186 RHEINGOLD, H. L. and HESS, E. H., 1957. 'The chick's "preference" for some visual properties of water': *J. comp. physiol. Psychol.* 50, 417–21.

187 RICHTER, C. P. and ECKERT, J. F., 1937. 'Increased calcium appetite of parathyroidectomized rats': *Endocrinology*, 21, 50–4.

188 RICHTER, F., 1954. 'Experiments to ascertain the causes of feather-eating in the domestic fowl': *Wld's. Poult. Congr. Xth.*, 258–62.

189 ROGERS, F. T., 1916. 'Contributions to the physiology of the stomach 39: The hunger mechanism of the pigeon and its relation to the central nervous system': *Am. J. Physiol.* 41, 555–70.

190 ROSS, S., GOLDSTEIN, I. and KAPPEL, S., 1962. 'Perceptual factors in eating behavior in chicks': *J. comp. physiol. Psychol.* 55, 240–1.

191 ROSSI, P. J., 1968. 'Adaptation and negative after-effect to lateral optical displacement in newly-hatched chicks': *Science* 160, 430–2.

192 SALOMON, A. L., LAZORCHECK, M. J. and SCHEIN, M. W., 1966. 'Effect of social dominance on individual crowing rates of cockerels': *J. comp. physiol. Psychol.* 61, 144–6.

193 SALZEN, E. A. and SLUCKIN, W., 1959. 'The incidence of the following response and the duration of responsiveness in domestic fowl': *Anim. Behav.* 7, 172–9.

194 SAUER, C. O., 1952. 'Agricultural origins and dispersals': *Am. Geog. Soc.*, New York.

195 SCHAIBLE, P. J., DAVIDSON, J. A. and BANDEMER, S. L., 1947. 'Cannibalism and feather-pecking in chicks as influenced by certain changes in a specific ration': *Poult. Sci.* 26, 651–6.

196 SCHALLER, G. B. and EMLEN, J. T., Jr., 1962. 'The ontogeny of avoidance behaviour in some precocial birds': *Anim. Behav.* 10, 370–81.

197 SCHIFFMAN, H. R. and WALK, R. D., 1963. 'Behavior on the visual cliff of monocular as compared to binocular chicks': *J. comp. physiol. Psychol.* 56, 1064–8.

198 SCHJELDERUP-EBBE, Th., 1922. 'Beiträge zur Social-psychologie des Haushuhns': *Z. Psychol.* 88, 225–52.

199 SCHJELDERUP-EBBE, Th., 1923. 'Weitere Beiträge zur social-und-individual Psychologie des Haushuhns': *Z. Psychol.* 92, 60–87.

200 SCHOLES, N. W., 1965. 'Detour learning and development in the chick': *J. comp. physiol. Psychol.* 60, 114–16.

201 SCHOOLLAND, J. B., 1942. 'Are there any innate behavior tendencies?': *Genet. Psychol. Monogr.* 25, 220–87.

202 SHINKMAN, P. G., 1963. 'Visual depth discrimination in day-old chicks': *J. comp. physiol. Psychol.* 56, 410–14.

203 SIEGEL, P. B. and GUHL, A. M., 1956. 'The measurement of some diurnal rhythms in the activity of White Leghorn cockerels': *Poult. Sci.* 35, 1340–5.

204 SIEGEL, H. S., 1959. 'Egg production characteristics and adrenal function in White Leghorns confined at different floor space levels': *Poult. Sci.* 38, 893–8.

205 SIEGEL, H. S. and SIEGEL, P. B., 1961. 'The relationship of social competition with endocrine weights and activity in male chickens': *Anim. Behav.* 9, 151–8.

206 SIEGEL, P. B., 1965. 'Genetics of behavior: selection for mating ability in chickens': *Genetics* 52, 1269–77.

207 SIREN, M. J., 1963. 'A factor preventing cannibalism in cockerels': *Life Sci.* 2, 120–4.

208 SKARD, A. G., 1937. 'Studies in the psychology of needs': *Acta. Psychol.* 2, 175–229.

209 SLUCKIN, W., 1964. 'Imprinting and early learning': Methuen, London.

210 SMITH, F. V., 1960. 'Towards a definition of the stimulus situation for the approach response of the domestic chick': *Anim. Behav.* 8, 197–200.

211 SMITH, F. V. and HOYES, P. A., 1961. 'Properties of the visual stimuli for the approach response in the domestic chick': *Anim. Behav.* 9, 159–66.

212 SMITH, W., 1957. 'Social "learning" in domestic chicks': *Behaviour* 11, 40–55.

213 SMITH, W. and HALE, E. B., 1959. 'Modification of social rank in the domestic chicken': *J. comp. physiol. Psychol.* 52, 373–5.

214 SPALDING, D., 1873. 'Instinct': *MacMillan's Magazine* 27, 282–93. Reprinted 1954 *Brit. J. anim. Behav.* 2, 2–11.

215 STADELMAN, W. J., 1958a. 'The effects of sounds of varying intensity on hatchability of chicken egg': *Poult. Sci.* 37, 166–9.

216 STADELMAN, W. J., 1958b. 'Observations with growing chickens on the effects of sounds of varying intensities': *Poult. Sci.* 37, 776–9.

217 STAVSKY, W. H. and PATTIE, F. A., 1930. 'Discrimination of direction of moving stimuli by chickens': *J. comp. Psychol.* 10, 317–23.

218 SZYMANSKI, J. S., 1918. 'Einige Bemerkungen über die biologische Bedeutung akustischer Reize': *Archiv. für Physiologie.* 171, 373–83.

219 TALLARICO, R., 1961. 'Studies of visual depth perception. III. Choice behavior of newly-hatched chicks on a visual cliff': *Percept. Mot. Skills* 12, 259–62.

220 TALLARICO, R. and FARRELL, W. M., 1964. 'Studies of visual depth perception: an effect of early experience on chicks on a visual cliff': *J. comp. physiol. Psychol.* 57, 94–6.

221 TEITELBAUM, P., 1961. 'Disturbances in feeding and drinking behaviour after hypothalamic lesions': *Neb. Symp. Motiv.* 39–69 Univ. Nebraska, Lincoln.

222 THORPE, W. H., 1963. 'Learning and instinct in animals'. Methuen, London.

223 THOMPSON, W. R. and DUBANOSKI, R. A., 1964. 'Early arousal and imprinting in chicks': *Science* 143, 1187–8.

224 TINDELL, D. and CRAIG, J. V., 1959. 'Effects of social competition on laying house performance in the chicken': *Poult. Sci.* 38, 95–105.

225 TINDELL, D. and CRAIG, J. V., 1960. 'Genetic variation in social aggressiveness and competition effects between sire families in small flocks of chickens': *Poult. Sci.* 39, 1318–20.

226 TOLMAN, C. W. and WILSON, G. F., 1965. 'Social feeding in domestic chicks': *Anim. Behav.* 13, 134–42.

227 TOLMAN, C. W., 1967. 'The effect of tapping sounds on feeding behaviour of domestic chicks': *Anim. Behav.* 15, 145–8.

228 TRETYAKOV, N. P., 1953. 'Inkubatsiya': Gosudarstvennoye Izdatyelstvo Syelskochozyastvenno Litteraturi, Moscow.

229 TUCKER, D., 1965. 'Electrophysiological evidence for olfactory function in birds': *Nature Lond.* 207, 34–6.

230 UPP, C. W., 1927. 'Preferential mating in fowls': *Poult. Sci.* 7, 225–32.

231 VAN TIENHOVEN, A., 1961. 'Endocrinology of reproduction in birds', in 'Sex and Internal Secretion': Ed. Young, W. C., Wilkins, W., Baltimore.

232 VAN TIENHOVEN, A. and COLE, R. K., 1962. 'Endocrine disturbances on obese chickens': *Anat. Rec.* 142, 111–22.

233 VIDAL, J-M, 1967. 'Contribution a l'étude des comportement sexuals precoces des poussins de *Gallus domesticus*': *C. R. Acad. Sc. Paris D* 264, 2392–4.

234 VINCE, M. A. 1964. 'Synchronization of hatching in American Bobwhite Quail (*Colinus virginianus*)': *Nature Lond.* 203, 1192–3.

235 VINCE, M. A., 1964. 'Social Facilitation of hatching in the Bobwhite Quail': *Anim. Behav.* 12, 531–4.

236 VON HOLST, E. and VON SAINT PAUL, U., 1960. 'Vom Wirkungsgefüge der Triebe': *Naturwissenschaften* 18, 409–22.

237 VON HOLST, E. and VON SAINT PAUL, U., 1963. 'On the functional organization of drives': *Anim. Behav.* 11, 1–20.

238 VON SAALFELD, E. F., 1936. 'Untersuchungen über das Hacheln bei Tauben': *Z. vergl. Physiol.* 23, 727–43.

239 WALLS, G. L., 1942. 'The vertebrate eye': Cranbrook Institute of Science Bull. No. 19.

240 WARREN, J. M., BROOKSHIRE, K. H., BALL, G. G. and REYNOLDS, D. V., 1960. 'Reversal learning by White Leghorn chicks': *J. comp. physiol. Psychol.* 53, 371–5.

241 WATSON, J. B., 1916. 'The place of the conditioned-reflex in psychology': *Psychol. Rev.* 23, 89–116.

242 WILSON, W. O. and WOODARD, A. E., 1958. 'Egg production of chickens kept in darkness': *Poult. Sci.* 37, 1054–7.

243 WINKELMANN, R. K. and MYERS, T. T., 1961. 'The histochemistry. and morphology of the cutaneous sensory end-organs of the chicken': *J. comp. Neurol.* 117, 27–36.

244 WINSLOW, C. N., 1933. 'Visual illusions in the chick': *Archiv. Psychol.* 153, 5–82.

Bibliography

245 WOLFE, J. B. and KAPLON, M. D. 1941. 'Effect of amount of reward and consummative activity on learning in chickens': *J. comp. Psychol.* 31, 353–62.

246 WOLFF, H. G., 1960. 'Stressors as a cause of disease in man': in Stress and psychiatric disorder. Ed. Tanner, J. M. Blackwell, Oxford.

247 WOOD-GUSH, D. G. M., 1954a. 'The courtship of the Brown Leghorn Cock': *Brit. J. Anim. Behav.* 2, 95–102.

248 WOOD-GUSH, D. G. M., 1954b. 'Observations on the nesting habits of Brown Leghorn hens': *Wlds. Poult. Congr.* Xth 187–92.

249 WOOD-GUSH, D. G. M., 1955. 'The behaviour of the domestic chicken: review of the literature': *Brit. J. anim. Behav.* 3, 81–110.

250 WOOD-GUSH, D. G. M., 1956. 'The agonistic and courtship behaviour of the Brown Leghorn cock': *Brit. J. Anim. Behav.* 4, 133–42.

251 WOOD-GUSH, D. G. M., 1957. 'Aggression and sexual activity in the Brown Leghorn cock': *Brit. J. Anim. Behav.* 5, 1–6.

252 WOOD-GUSH, D. G. M., 1958a. 'The effect of experience on the mating behaviour of the domestic cock': *Anim. Behav.* 6, 68–71.

253 WOOD-GUSH, D. G. M., 1958b. 'Fecundity and sexual receptivity in the Brown Leghorn female': *Poult. Sci.* 37, 30–3.

254 WOOD-GUSH, D. G. M., 1959a. 'A history of the domestic chicken from antiquity to the 19th Century': *Poult. Sci.* 38, 321–6.

255 WOOD-GUSH, D. G. M., 1959b. 'Time-lapse photography: a technique for studying diurnal rhythms': *Physiol. Zoöl.* 32, 272–83.

256 WOOD-GUSH, D. G. M., 1960. 'A study of sex drive of two strains of cockerels through three generations': *Anim. Behav.* 8, 43–53.

257 WOOD-GUSH, D. G. M., 1963a. 'The control of the nesting behaviour of the domestic hen I. The role of the oviduct': *Anim. Behav.* 11, 293–9.

258 WOOD-GUSH, D. G. M., 1963b. 'The relationship between hormonally-induced sexual behaviour in male chicks and their adult sexual behaviour': *Anim. Behav.* 11, 400–2.

259 WOOD-GUSH, D. G. M. and GILBERT, A. B., 1964. 'The control of the nesting behaviour of the domestic hen II. The role of the ovary': *Anim. Behav.* 12, 451–3.

260 WOOD-GUSH, D. G. M., 1965. 'The social organization of domestic bird communities': *Symp. Zool. Soc. Lond.* 14, 219–31.

261 WOOD-GUSH, D. G. M. and KARE, M. R., 1966. 'The behaviour of calcium-deficient chickens': *Brit. J. Poult. Sci.* 7, 285–90.

262 WOOD-GUSH, D. G. M. and GOWER, D. M., 1968. 'Studies on the motivation in the feeding behaviour of the domestic cock': *Anim. Behav.* 16, 101–7.

263 WOOD-GUSH, D. G. M. and GILBERT, A. B., 1968. 'Observations on the laying behaviour of hens in battery cages': *Brit. Poult. Sci.* (In press.)

264 WOOD-GUSH, D. G. M. and OSBORNE, R., 1956. 'A study of differences in the sex drive of cockerels': *Anim. Behav.* 4, 102–10.

265 ZAWADOVSKY, B. M. and ROCHLINA, M. L., 1929. 'Bedingte Reflexe bei normalen und hyperthyreoidisierten Hühnern': *Z. vergl. Physiol.* 9, 114–44.

266 ZIELINSKI, K., 1960. 'Studies on higher nervous activity in chicken. I. The effect of age on conditioned alimentary excitatory and inhibitory reflexes': *Act. Biol. exp. Vars.* 20, 65–77.

Index

Index